MARINE
WILDLIFE
of
ATLANTIC EUROPE

MARINE
WILDLIFE
of
ATLANTIC EUROPE

by Amanda Young
Photography by Paul Kay

IMMEL
Publishing

Marine Wildlife of Atlantic Europe is published by
Immel Publishing

Copyright © 1994 text: Amanda Young
Copyright © 1994 photographs: Paul Graham Kay
Copyright © 1994 layout and design: Immel
Publishing Ltd

Design and cover: Jane Stark
Typesetting: Johan Hofsteenge
Production: Paula Casey-Vine

Cataloguing in Publication Data
A CIP catalogue record for this book is available
from the British Library

ISBN 0 907151 81 7

Immel Publishing Ltd
20 Berkeley St., Berkeley Square, London W1X 5 AE
Tel: 071 491 1799. Fax: 071 493 5524

CONTENTS

ACKNOWLEDGEMENTS   6

FOREWORD   7

INTRODUCTION   9

PART 1 : HABITATS   11

PART 2 : ANIMALS   37

        Sponges   40
        Jellyfish and anemones   44
        Molluscs   66
        Crabs and lobsters   73
        Starfish and urchins   103
        Fish   122

GLOSSARY   183

INDEX   186

ACKNOWLEDGEMENTS

The photographs in this book were taken at a large number of locations and particular thanks must be given to Matt Murphy (Director of the Sherkin Is. Marine Station) and Anglesey Sea Zoo for their cooperation and assistance. Photographs of fossil material were taken with the kind permission of Stone Science.

In writing this book we are indebted to the numerous scientists who have, by their study and observation, provided the knowledge on which we have drawn so heavily, particularly to Dr Rod Jones, Adrian Brooks and Joe Foxcroft for their help and support.

Jewel anemones and boring sponge: not only are some creatures exquisitely delicate, they are often surprisingly colourful.

FOREWORD

T hink of the word 'holiday', and immediately 'seaside' springs to mind. Yes, it is a wonderful and exciting place where you can turn your back on the problems of a workaday life and seek new hope in new horizons. Well, you could in my young days.

Yes, I was one of the bucket and spade brigade who tormented the denizens of the rockpools as I made my first dabblings into natural history.

Today much has changed: sewage, plastic rubbish, oil-soaked birds and dead seals litter the beachscapes, and overfishing, pollution and enrichment threaten the living balance of the seas. Despite this, more and more people are taking to the waters with fins, masks, snorkels and scuba gear, eager to learn before it is too late. Here is a handbook of the marine wildlife of Atlantic Europe, written and photographed by two experts who still have that wonder of childhood experience within their grasp. It describes habitats and their inhabitants in a manner that will educate, entertain and enthral.

Let this book guide you along our precious coastline so that you may become experts yourselves and so champions of conservation before it is too late.

Thank you for caring.

David Bellamy
Bedbur

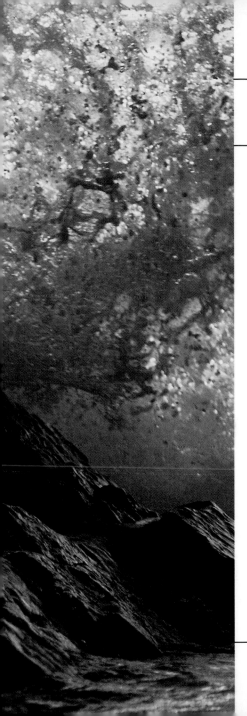

INTRODUCTION

Many books have been written about the inhabitants of the shore and shallow seas. The aim of this book is to complement those of a more general nature by focusing closely on individual facets of the marine world and its inhabitants.

The first part of the book aims to give the reader an insight into the wide variety of environments found around Britain's coast: from vast expanses of golden sandy beach to tall cliffs pounded by huge waves which have rolled their way across the Atlantic Ocean; from the tranquillity of sea lochs on a calm winter's morning to the bustling activity in a spring rockpool, where fish and prawns dart about among the numerous seaweeds and encrusting animals that share their home. All are linked in their ability to provide homes for marine animals.

In the second part of the book individual fish and marine invertebrates are discussed in more detail. No one book can cover all the tremendous diversity of our coastal wildlife, and no apologies are made for missing creatures or for the selective nature of the text. Rather, we offer an insight into some of the more fascinating or more often encountered animals in the inshore waters around our coasts. Looking at this wildlife, we are constantly surprised by both their beauty and their intricate lifestyles, and hope that in this volume we can share some of this enjoyment with you.

Calm surface conditions can disguise fast-running tidal currents.

HABITATS

Right: Rocks may appear superficially barren, but actually provide the most prolific habitats.

Facing page: A period of very calm weather occasionally results in exceptional water clarity.

The rocks of rocky shores all offer a firm substrate even though they can vary in age by as much as three billion years. They will also vary a great deal in strength, depending on the local geology. Some, like chalk, are friable in the small scale while others, such as carboniferous limestone, may disintegrate in jointed blocks, and yet others are massive granites.

The types of plants and animals that live on the rocky shore may vary considerably from place to place. The particular selection found at any one site will depend on the rock type and the degree of shelter or exposure to wind and waves, together with the aspect and angle of the rock face. Some of the animals that are found will attach to any type of rock and simply use it as a base from which to capture the plentiful array of planktonic organisms that drift past on the current. Others have very specific requirements. One such example is the piddock, a bivalve mollusc, which drills and bores its way into soft sandstone and limestone. It effectively forms its own grave, because as it grows and expands its home, it becomes too big to leave by the narrow entrance to the sea.

On sheltered rocky shores there may be so many plants and animals attached to the surface that the rock face is totally obscured. Numerous blennies, butterfish, worm pipefish, crabs and anemones can be found hiding or making their home beneath the dense expanse of abundant brown fucoid weed. These plants both protect the rock from the erosive power of the waves and provide enough shelter and moisture – together with insulation from extreme cold – for the animals to survive in air for the few hours during which the tide is out.

In contrast to this, where the rocks are exposed to the constant pounding of waves and the frequent

ROCKY SHORES

Little life can survive where the rock surface is too soft to withstand pounding waves.

violence of storms, far less seaweed is able to attach and grow. In its place barnacles smother the rock, providing food for predatory dog whelks and crevices in which periwinkles hide. Mussels are also able to withstand such harsh conditions on rocky promontories. At times thousands huddle closely together, reducing each other's exposure to the waves. They hang on tightly to one another and to the rocks by up to 300 byssus threads. It is interesting to note that mussels growing in exposed places rarely exceed 2cm in length, whereas in a sheltered sea loch they may be over 10cm long.

In some areas of the coastline cliffs form almost vertical walls of rock plummeting down into the sea, often to some considerable depth. Waves rolling in across the ocean hit the cliffs repeatedly with a force of up to 25 tonnes per square metre. Under such conditions only the most hardy of creatures can survive. Indeed, if the cliff is without cracks and crevices or is of soft material that is constantly crumbling away, it may be totally lifeless.

Top left: Zones of barnacles, red algae and kelp show at low spring tides.

Bottom left: Sponges, starfish and feather stars all live together, although not necessarily in harmony!

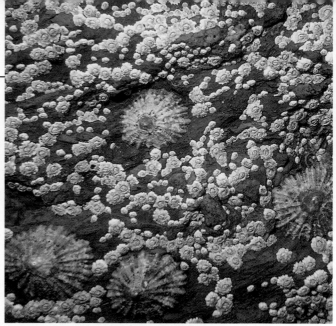

Above: In this gully at low-tide, the clear water shows kelp well below the low-tide limit.

Above: Shellfish adhere to rock surfaces in any area where they can withstand the wave action.

Right: Rockpools provide habitats for a variety of life.

Facing page: Weed-filled rockpools provide shelter for many animals.

Rockpools form because of cracking or differential weathering of the underlying shore material. Each time the tide goes out it leaves behind a freshly filled pool of seawater together with a variety of animals that remain stranded until the tide returns. Although these creatures will not suffer from desiccation like those on the exposed shore, it is by no means an easy environment. Depending on the time of year conditions may vary widely. In winter the small amount of water will become far cooler, and in summer far hotter, than does the surrounding sea. The summer sun may also cause considerable evaporation so the salt content will rise. Conversely, if it rains heavily the salt will be diluted. In addition, although seaweeds release oxygen by day they use it up at night and animals may suffocate. These problems are most severe high on the shore, which is why such pools are relatively barren in comparison to those lower down.

Well-lit pools near low-water mark may contain almost any of the marine wildlife commonly associated with the coast. Some organisms do particularly well and the pink encrusting seaweed *Corallina officinalis* frequently coats the rock together with sponges and other colonial animals. Seaweeds give shelter and protection to young crabs and prawns. Limpets graze the algae from the rocks and the surprisingly mobile beadlet anemone awaits its chance to catch its unsuspecting prey. In the spring sea slugs such as sea lemons which eat the breadcrumb sponge, and sea hares, which browse on seaweeds, move from deeper water into intertidal pools where they lay their eggs.

A number of intertidal fish are frequently trapped by the tide. Nearly all have mottled brown/green coloration which helps to camouflage them amongst the weed and debris of the pool. Most are without scales, so they can slip into the safety of nooks and crannies; others, like gobies, have fan-shaped suckers with which they can grip the rocks. All these creatures must await the incoming tide for refreshed seawater, a new batch of food or their liberty.

ROCKPOOLS

Left: Seaweeds grow on each other, producing a jungle-like effect.

Below: Kelp beds not only look like forests, but they can be equally dense and almost impenetrable.

*Top: Kelp is not confined
only to dense forest
growths, but can also be
found in the shallows
amongst other plants.*

*Bottom: The star
ascidian is often found
around kelp holdfasts.*

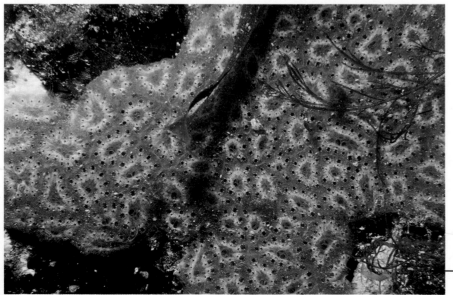

One of the most exciting underwater
landscapes is that of the kelp forest, huge
areas of which are found in shallow rocky
regions around much of Britain's coastline.
Most species of kelp are perennials and some live for
over ten years. They grow at and below the low-tide
level and are therefore usually more familiar to people
who scuba dive than to those who explore the
seashore. Around Britain most types of kelp are of the
Laminaria family, the largest of which, *Laminaria
saccharina*, the sugar kelp, can grow to as much as
4m in length.

Unlike a land plant there are no roots; instead the
kelp is attached by means of a holdfast with which it
grips the rock. From the holdfast a thick stipe lifts

KELP FORESTS

Right: Blue-rayed limpets are commonly found on kelp.

Below: Thick kelp may be exposed at low spring tides.

aloft a leathery blade. This blade may be a single unit or divided into fronds; either way it provides an ideal surface upon which a variety of plants and colonial animals, in particular the sea mat, *Membranipora membranacea*, settle and grow. If this covering becomes too thick the area of the blade exposed to light and able to photosynthesize the plant's food is reduced. The pattern of growth of the seaweeds is variable. *Laminaria digitata*, oarweed, is usually the dominant kelp in the forest because of its high growth rate, which continues from spring to autumn. In other kelp species growth tends to slow or stop by late summer.

Kelp forests are extremely productive, and they are the basis of a complex marine ecosystem. Fish are the main predators and wrasse are the most dominant species. The fronds are so long and the stipe so thick that in the Second World War submarines were able to hide, undetected by enemy sonar, in the kelp beds off north-west Scotland. For thousands of years, many species of fish have done likewise. Most are well camouflaged so that they look like or are coloured to match the weed, and so escape predators.

Other animals have a more selective approach. The herbivorous blue-rayed limpet, *Helcion pellucidum*, always prefers to migrate on to laminarian seaweeds when it has reached 4mm or so in size. It will then graze its way down to the safety of the holdfast where it takes up residence. Unfortunately, the constant grazing of the holdfast by the limpet may weaken the attachment to the point at which both plant and animal are swept away by a storm.

Pebbles provide one of the most difficult habitats for marine creatures to survive in, and extensive pebble beaches are relatively barren areas.

Shingle may either form a complete beach, such as the 29km stretch of Chesil Beach, which links the Isle of Porland to Dorset, or occur as a wide band above a sandy beach as at Newgale in Dyfed and most of the West Sussex coast. Shingle consists of rocks and boulders that are constantly rounded and ground down by the pounding of the waves. Sand remains moist because water is held between the grains by capillary action, but on this type of shore the boulders are large and the gaps between them too big for any moisture to be retained as the tide goes out. Because of this they quickly dry out and are unable to provide adequate shelter, moisture and oxygen to any marine wildlife that may be trapped in the gaps.

The larger the pebbles are, the greater the slope of the beach can be. At times beaches are thrown into a series of steep banks and terraces at right angles to the dominant wave action. As the waves break obliquely onto the shore, the pebbles are thrown to one side. This results in the material moving in the direction of the prevailing wind and is called 'long shore drift'.

The constant movement of the rocks and boulders is sufficient to both crush any animals trapped by the shingle and knock away any newly settled plant spores or larval animals. As a result a shingle beach is the most barren and arid of all British intertidal environments.

SHINGLE SHORES

SANDY SHORES

Below: Sand produces a characteristic, flat shorescape.

Right: Although superficially barren, sand provides a home for many creatures such as the sea potato, Echinocardium cordatum (inset).

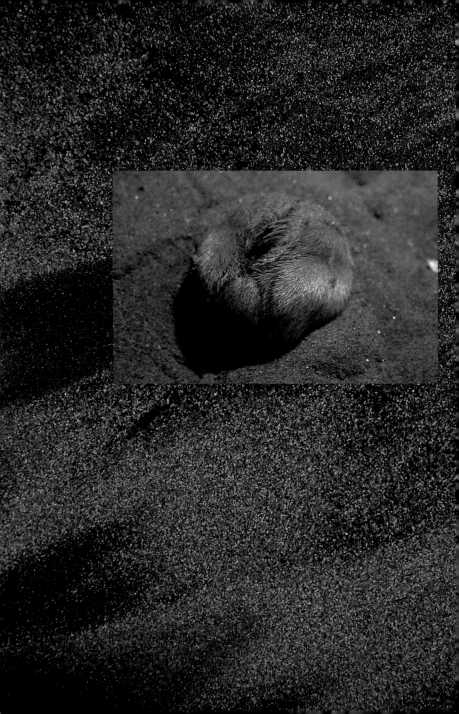

Rippled sand betrays its inhabitants in the form of so-called worm casts.

G ently sloping sandy beaches are found all around the country in bays between rocky headlands and at the mouths of estuaries, sometimes as sand spits. They occur anywhere, in fact, where the tides and currents may have carried and deposited the fine material.

Sand is made from the erosion of rocks and seashells. The grains are graded into four size catagories: coarse (0.5–2.0mm diameter), medium (0.25–0.5mm), fine (0.125–0.25mm) and very fine (0.063–0.125mm). The material is hard, often containing a high proportion of quartz, and durable enough to withstand the constant movement on and along the beach without being further ground down into mud. In between each small irregular grain there is just enough space for water to be held by capillary action. This seawater contains enough oxygen to sustain life. Although the surface of the beach may be hot and dry, a little digging will show that not far down it is cooler and damp. There may be a difference of 7°C between the temperature of the surface and that of the material 15cm below.

On a hot summer day when the tide is out, the sandy seashore is apparently lifeless – except perhaps for hordes of people with their picnics, buckets and spades! No seaweeds will be found on such a beach because they cannot survive and grow on the mobile sand. There is a variety of animals, however, all adapted to burrowing, that do survive well in the cool, dark environment beneath the surface. Most must wait for the tide to flood in before they can feed. Bivalve molluscs such as the razor shell and cockle live permanently buried. They extend their siphons above the surface of the sand and filter plankton from the sea. Other members of the same family, tellin for example, are deposit feeders and collect food by groping about over the surface with their siphons. Animals like shore crabs emerge and scavenge for anything that is edible, be it a left over sandwich or a dead fish! Some animals, such as the sea potato, are not so dependent on the state of the tide. They get their food by swallowing the sand and digesting the organic material attached to it.

SANDY SHORES

Sea grass is one of the few marine plants with roots and can be found in sandy estuaries and bays.

Worm casts can be found equally as well underwater, showing the presence of hidden creatures.

Unlike a superficially barren sandy beach, the sandy seabed can be a hive of activity. However, beneath the sea the grains of sand are still moved constantly by both waves and current and can build up into large mobile banks, such as the Goodwin Sands in the English Channel, which are exposed only on equinoctial spring tides. The shifting nature of the sand means that few plants, other than the sea grass *Zostera marina* which lives in sheltered estuaries and bays, can survive.

Fresh seawater, together with the vital oxygen it carries, penetrates deeper into sand than mud. Therefore different types of animals may survive, depending on the organic content, the amount of mud and the degree of sorting of the sand. Those sands which are clean and well graded provide a larger area in which to hide or live. Feeding and other activities are not limited by the time of high tide on the seabed, as they are on the shore so, irrespective of the time of day, bivalves can feed through their siphons. Sand masons *(Lanice conchilega)*, peacock worms *(Sabella pavonina)* and other worms extend their delicate siphons and gills. Hermit crabs scuttle over the surface looking for a meal at night and shrimps emerge from their daytime burrows in their thousands to patrol the sand in search of anything edible.

Many of the animals remain hidden in the sand during the day to avoid predation and then emerge to feed at night. Young flatfish are an example. They are able both to change their colour to camouflage themselves and to burrow under the sand with only an eye protruding to keep watch. Once night falls they migrate inshore to feed on the plentiful shrimps and other small invertebrates.

Large shoals of sand eel, an important source of food to many other fish and seabirds, dash in and out of the sand, their arrow-shaped bodies plunging between the grains like darts. Weever fish are another typical example of sand-dwelling fish. In shallow coastal waters they lie buried beneath the sand with only their eyes and poisonous dorsal fin rays projecting. Well protected, well hidden and constantly alert, they run little risk of being eaten themselves.

SANDY SEABEDS

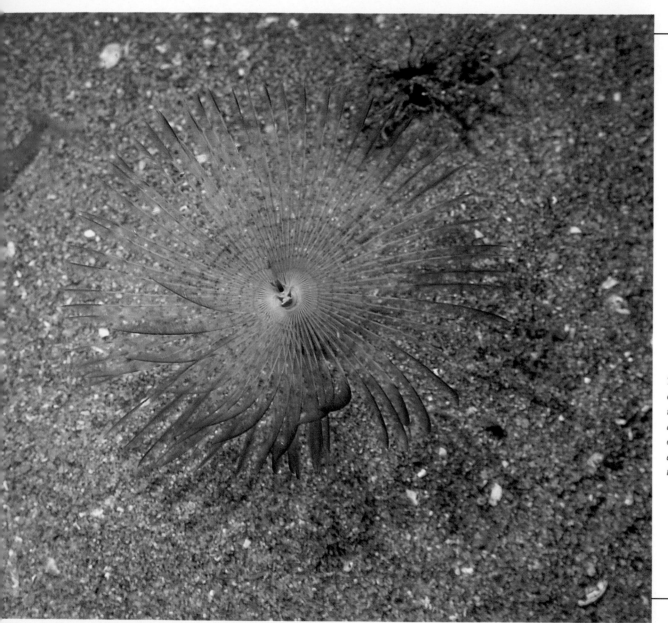

The peacock worm is an example of those worms which show themselves above the sand, relying on their ability to disappear quickly if threatened.

Left: Mud flats in the Dwyryd estuary, North Wales, at twilight.

M ud is made up from either firm, consolidated or loose, flocculent particles of silt (0.002–0.063mm in diameter) and clay (0.00006–0.002mm in diameter), mixed with rotting organic material. Mud flats are extremely productive and in many areas around Britain they cover huge areas. Examples include parts of the Wash, the Thames estuary, Solway Firth and Morecambe Bay. In the large estuary of the River Severn an interesting phenomenon occurs: pools of loose fluid mud develop during the neap tides but are swept away during the spring tides.

The mud found around our coasts comes from either the sea floor, erosion of the coast or material derived from river water. Mud can only accumulate in sheltered areas where the currents are so slow that the fine particles are able to fall out of suspension. This is most common in hollows in the sea floor, in wide estuaries, or in the lee of a spit or barrier. In such places the mud is often mixed with quantities of organic material and together they form a glutinous material. The surface may contain some oxygen but beneath is a black sulphide-rich layer. As it decomposes it releases hydrogen sulphide, which smells of rotting eggs, and methane. In the open sea the amounts released cause no problem. However in certain places, such as beneath a commercial salmon farm where the waste food falls to the seabed and

rots, hydrogen sulphide may bubble up out of the sediment and can impair water quality. The only creatures that manage to survive in this type of mud are those, like worms, which dig a U-shaped tube through which they pump down fresh oxygenated seawater. Mud such as this also contains numerous bacteria which do not need oxygen to survive and as a result find the decaying plants and animals a rich source of food. They in turn are fed on by burrowing animals of increasing size, including the iridescent sea mouse *(Aphrodite)* and sea cucumber *(Holothuria forskali)*.

On the surface of the mud there are often large mats of algae. Mullet swim over the surface skimming the algae off with their wide, soft mouths. However, it is the vast profusion of tiny snails *(Hydrobia ulvae)* – up to 15,000 per square metre – which browse on the microscopic seaweeds, attract the shoals of fish and large flocks of migrating wading birds, especially shelduck, that are so characteristic of a British mud flat.

Left: Close-up of a muddy shore.

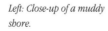

MUDDY SHORES

Facing page: Salt marshes often line wide estuaries.

Left: The spiky perennial grass, Spartina anglica *speeds up the establishment of salt marshes in many areas.*

An estuary is an embayed area in which land, river and sea meet together. Owing to the sheltered nature of most estuaries, there is usually only weak wave action, but the 'funnel' effect can cause dangerously strong tidal currents. There are also varying amounts of freshwater from the river, the twice-daily tides and seasonal temperature changes (remember, freshwater freezes at a higher temperature than seawater). The mass of the sea is relatively stable in terms of temperature and salt content and any changes that do take place occur gradually. In an estuary such changes are going on all the time. This means that it is an often hostile environment in which few animals can survive.

Most animals are able to live either in freshwater or in saltwater but would die if moved from one to the other. Only a few species such as mullet, salmon, flatfish and sticklebacks are able to make the change. As a result they are able to enjoy the advantage of feeding on invertebrate animals in the estuary with little competition from other species. Although the number of species found in an estuary is usually smaller than in most marine environments, because they are able to establish themselves with little competition each type can be found in huge numbers.

Salt marshes are frequently found in association with the landward side of mud flats and particularly in estuaries. These, at times extensive, areas trap sediment which is then stabilized by the growth of salt-tolerant sea grasses, herbs and dwarf shrubs. Material carried into the salt marsh by the tide is filtered out and remains lodged between the plants. As it accumulates banks build up, reclaiming land, and in between little creeks develop together with small basins, called pans.

In the safe, sheltered creeks, where the water warms quickly with the approach of summer, a variety of young fish grow and develop. Such areas are of particular importance to the young of the commercially valuable sea bass. Opossum shrimps *(Neomysis integer)*, known also as mysids, are particularly abundant in the creeks and provide an important source of food to the young fish.

In many areas of Britain today the establishment of a salt marsh is speeded up by a tough, spiky perennial grass called *Spartina anglica*. It has evolved from a vigorous hybrid cross between British and American *Spartina* plants. Ever since its emergence in Southampton water in the 1870s, it has spread rapidly around the coastline, changing the structure and altering the development of our salt marshes. The plant is able to colonize areas where other salt marsh plants cannot survive and now forms up to 50 per cent of the salt marsh in some estuaries.

The organic debris produced by salt marshes, which are some of the most productive ecosystems on earth, is incorporated into the surrounding sediments, raising their level of productivity. Salt marshes only normally exist in Britain above the mean high water neap tide level. Some salt marsh species such as *Phragmites* are found in areas of lower salinity on the upper shore, away from a more marine environment.

ESTUARIES AND SALT MARSHES

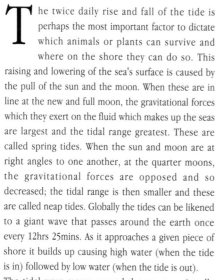

The twice daily rise and fall of the tide is perhaps the most important factor to dictate which animals or plants can survive and where on the shore they can do so. This raising and lowering of the sea's surface is caused by the pull of the sun and the moon. When these are in line at the new and full moon, the gravitational forces which they exert on the fluid which makes up the seas are largest and the tidal range greatest. These are called spring tides. When the sun and moon are at right angles to one another, at the quarter moons, the gravitational forces are opposed and so decreased; the tidal range is then smaller and these are called neap tides. Globally the tides can be likened to a giant wave that passes around the earth once every 12hrs 25mins. As it approaches a given piece of shore it builds up causing high water (when the tide is in) followed by low water (when the tide is out). The tidal wave moves around the equator, virtually

undetectable, at a speed of about 1,000mph. Only when it reaches land are the resulting strong currents seen. In Wales, Bardsey Island is called Ynys Enlli – Island of Currents – and between it and the mainland currents reach speeds of 8 knots. Tides are also seen to great effect in estuaries such as those of the River Severn and River Dee. In these, the shape of the estuary is such that the water is squeezed together and eventually forms a tidal bore. This is one large wave which rushes upstream at between 10 and 20mph in a vertical wall of water that can reach as much as 3m high.

No matter what the range, the effect of the tides on the shore is to expose part of the seabed to air (or

part of the land to the sea!). Plants and animals normally living beneath the surface of the sea in a fairly uniform environment have to be able to adapt to this exposure. They may suffer desiccation from the sun and the wind. They may experience changes in salinity from freshwater streams and rain and they may have to endure extreme cold on frosty winter nights or predation from land animals.

These and other factors have meant that seashore life has had to become well adapted to its fluctuating environment. Each type of animal or plant may only settle and survive in a small tidal band or zone along the coastline. Because of this, rocky shores in particular look superficially similar all over the world.

TIDES

Above: Storms are fierce and destructive above the waves and can cause damage to the sea's inhabitants.

Right: Gentle, rolling waves indicate a shallow sloping seabed.

Waves, gently lapping or fiercely pounding against the coast reflect the constant motion of the sea. When you watch waves rolling, it looks as if the water is surging along. In fact, in the open ocean the water at any one point usually moves in a circular path so that there is little if any forward movement. It is energy, not material, that is transported and waves are an important means by which energy is moved around the oceans. Only as waves move into shallow water are they distorted by friction against the bottom so that they become steeper and steeper until they eventually break and crash on to the shore.

Waves are generated by the friction of the wind against the sea surface. The length of time the wind blows, the distance it is in contact with the sea surface (the fetch) and the speed at which it blows all influence how big the waves will be. The influence of wind strength on wave height is given by the Beaufort Scale. This table has thirteen categories, from 0 – Calm – to 12 – Hurricane and describes the wind speed, the visual state of the sea surface and the measured wave height. To the west of the country, where the wind has come from across the Atlantic Ocean, waves in the open ocean may be up to 35m in height. In contrast, on eastern coasts, even after

WAVES

prolonged gales, they rarely reach 20m. Moving down through the water column the influence of the waves decreases, until at a depth of about twice the wave height it ceases to be noticeable. Hence although divers are often bounced about by waves near the surface, they can soon escape to quieter water at a lower depth.

Occasionally, just as a pebble thrown into a pond causes ripples, an earthquake, volcanic explosion or submarine landslide may cause shock waves. These waves are called tsunami (they are also sometimes known as tidal waves, although they have nothing to do with tides) and travel at about 500mph. While they can have a devastating effect close to their source, by the time they have reached the British Isles they tend to be small and generally pass unnoticed.

Throughout the year the weather, and with it the power and effect of the waves, varies enormously. During the more tranquil summer months sand is gradually accumulated on beaches and plants and animals colonize the rocky shore. Then, during the severe storms that usually accompany autumn, there can be a complete change in the shorescape. As waves approach the shore their energy becomes concentrated onto headlands, which is why the largest waves are seen pounding the peninsulas of Europe's western coasts. Under these conditions all but the most tenacious organisms are knocked off and swept away. The large waves also sweep the sand away from the beaches and back out to sea on to underwater slopes, to form winter sand bars.

In sheltered estuaries and sea lochs the short fetch and limited water depth means that waves always remain small and organisms such as mussels can grow far larger than on more exposed open coasts. The degree to which a coast is defined as exposed depends on two factors: whether or not it faces into the direction of the prevailing wind, and the extent to which it is protected by offshore islands, sandbars and the like. Because the prevailing winds are from the south-west, the coastline to the west is more exposed and has a wider splash zone than that to the east, which gives a larger area in which marine and terrestrial organisms overlap.

Left: Waves break as they are distorted by friction against the sea bottom.

Facing page: Atlantic swells are powerful even on calm summer days.

Normally animals living on the shore cope well with the changing seasons and are un-affected by the occasional hot or cold day.

In summer open sea temperatures may reach from 13°C in the north of Scotland to 18°C around Cornwall. At times some shallow sheltered places like the Wash may even become a few degrees warmer. The warm spring weather and increasing day length herald the plankton bloom and the onset of courtship and breeding in most coastal wildlife. If there is an unusually prolonged spell of hot weather, some creatures will start to suffer and react. Mobile animals such as the viviparous blenny (*Zoarces viviparus*) move offshore. Many others are able to burrow deeper into the sediment, and only those which cannot migrate are left exposed to the extreme of a heatwave. Even these are surprisingly well able to endure such conditions. Those with shells, like the barnacle *Chthamalus stellatus*, which has been known to survive temperatures of 36°C (around human body temperature), close up tightly until the tide comes in.

It is in winter that severe weather causes the highest mortality among marine wildlife. Although the temperature of the open sea rarely drops below 5°C, during prolonged anticyclonic weather the air temperature may remain well below zero for a period of days. Groundwater seeping on to beaches freezes into beautiful organ-pipe formations. In estuaries and other low-salinity areas the sea itself can freeze over.

When this happens the animals below may die, not only from the cold, but also from lack of oxygen. Whereas fish are able to move offshore to deeper, warmer water, bivalve molluscs, including cockles and razor shells, at times die in their thousands. Happily, although a population from a particular location may be wiped out during a bad winter, recolonization rapidly takes place the following year.

TEMPERATURE EXTREMES

Shellfish have been on earth for millions of years.

ANIMALS

Left: Trilobites, long since extinct, were early marine creatures.

Facing page: Some fossils are clearly ancestors of present-day animals.

It has taken over 600 million years for the life that we find in the seas around Britain today to evolve. During that time numerous different species have arisen. Most have then, after a varying period of time, become extinct. They were unable to adapt rapidly enough to survive in an ever-changing environment.

Some of the earliest evidence of this life can be seen in the Cambrian rocks of Wales where the fossil remains of organisms such as trilobites and primitive gastropods are found. These animals inhabited the water and sediments of the prehistoric ocean. Although there are no coral reefs around the coastline today, large areas of the country are composed of carboniferous limestone which has fine examples of both solitary and colonial corals. These reflect the fact that the seas at that time were much warmer than they are now. Giant ammonites, close relatives of the octopus and squid, can be found in the Jurassic and cretaceous rocks of southern England. Today on certain beaches on the south coast fossil shark teeth are found which have been washed out of soft Tertiary sediments.

During the last ice age sea levels were lowered by over 100m and it became possible to walk from France to Britain. When the ice retreated, and the water closed the land bridge once more, the new area of seabed would have been quicky colonized by many of the species we still find today.

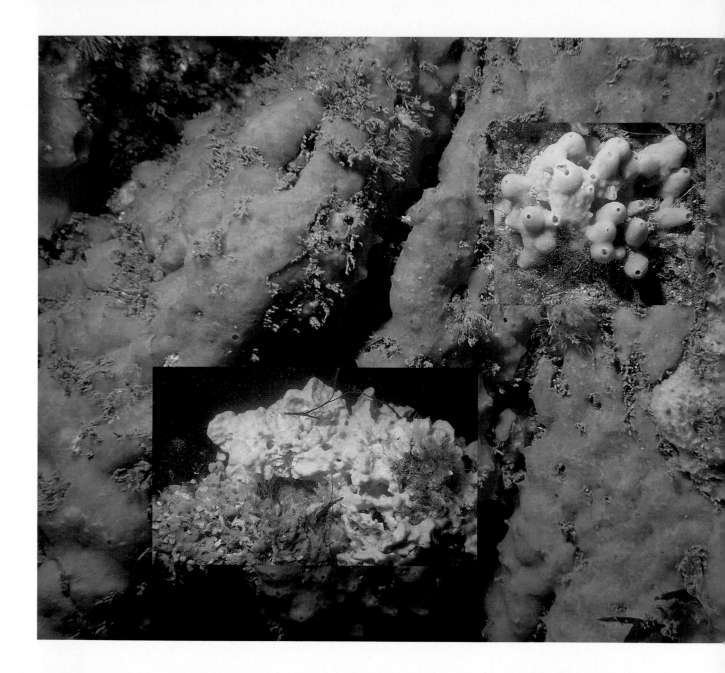

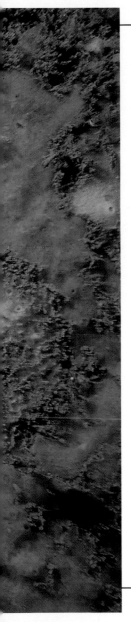

Facing page: Sponges can be abundant and colourful, but difficult to identify with certainty.

Far right: Some creatures such as these red sea squirts are superficially similar to sponges.

Sponges are, in terms of their structure and function, the simplest of multicellular animals. They are also very difficult to differentiate because of their superficial similarity to one another. It is often necessary to look first at their skeleton to find out whether it is made of spongin fibres or contains either silica or calcareous spicules (spines of varying shape). Once this has been done, the nature of the skeleton and the size, type and shape of the spicules (which can only properly be seen with the help of a microscope), can be used to make an accurate identification.

All sponges live attached to some sort of surface, usually stone, shell or wood, which they grow over or bore into. They prefer the damp, sheltered environment of gullies and rock overhangs, and can at times spread over extensive areas. In dark places they are usually dull grey or cream in colour but in areas that are well lit, they often support symbiotic algae which colour them various shades of green, orange or red. A close look will show that the basic body shape is that of an individual or a group of vases. Tiny cilia (hair-like structures) projecting from the cells that make up the animal cause a water current to flow in through the minute holes that are found all over the sponge and out again through a larger pore or pores on the surface.

For all their apparent simplicity, sponges display a huge variety in their body shape, form and colour. They are also famous for their ability to regenerate from damaged fragments, an important characteristic since they practise asexual as well as sexual reproduction. Some species even show degeneration and fragmentation during the winter when there is little food, followed by regeneration and growth during the abundant summer months.

Sponges play an important role in the marine environment. Some are seen to be in specific

SPONGES

BORING SPONGE	*Cliona celata*
Size:	Encrusting, may reach 1m across. Forms small holes in shell, each about 2mm in diameter.
Colour:	Yellow/olive green
Habitat:	Limestone rocks and shells; shore and shallow sea to 100m

associations. The sea orange is commonly found on the shell of a hermit crab, which enables it to be transported from one feeding area to another. After a time the sponge may completely dissolve away the shell and become the hermit crab's sole form of protection. Most sponges are used as food by a variety of other animals, such as sea slugs and sea urchins. In addition, they may also be an important place of refuge for small worms, crabs and other animals.

Above: One of the more easily identifiable sponges is the boring sponge, Cliona celata.

Left: Boring sponge with sea anemones.

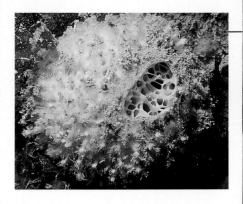

BREADCRUMB SPONGE
Halichondria panicea

Size:	Up to 20cm across and 2cm thick. Breaks easily.
Colour:	Olive green/brown
Habitat:	Rocks, seaweeds and shell; midshore and coastal sea to 10m

Halichondria bowerbanki

Size:	Spreading up to 20cm with long, finger-like extensions. Does not break easily.
Colour:	Cream/yellow
Habitat:	Rocks, boulders and wrecks; midshore to 30m

SEA ORANGE *Suberites domuncula*

Size:	Globe-shaped, up to 25cm diameter
Colour:	Orange or yellow/brown
Habitat:	Rocks and whelk shells inhabited by hermit crabs; shore and shallow water to 200m

Above left: Tethya aurantium, *while similar to the boring sponge, has only a single opening on top.*

Above right: The elephant's ear sponge is aptly named.

Below: Smaller elephant's ear sponges are often less irregular in shape than larger ones.

PURSE SPONGE *Grantia compressa*

Size:	Found singly or in groups each up to 5cm in height
Colour:	Cream or grey
Habitat:	Rocky seabed and among the red seaweed *Plumaria elegans*; low shore and shallow sea

ELEPHANT'S EAR SPONGE
Pachymatisma johnstonia

Size:	Up to 15cm high and 60cm across
Colour:	Grey outside and cream inside
Habitat:	Rocks and boulders; low shore and shallow sea

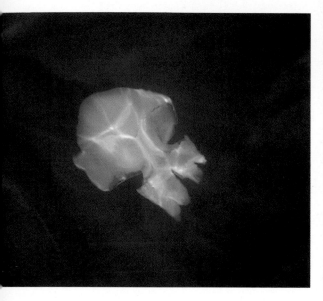

Above: An above-water view of the barrel jellyfish, Rhizostoma octopus.

Right: Barrel jellyfish are usually noticed on the beach where their size distinguishes them.

Facing page: The top of the compass jellyfish is similar to the old mariner's compass.

BARREL JELLYFISH	*Rhizostoma octopus*
Size:	Up to 90cm across bell
Colour:	Translucent blue/white/yellow, purple edge to bell
Habitat:	Open sea

COMPASS JELLYFISH	
Chrysaora hysoscella	
Size:	Up to 30cm across bell
Colour:	Transparent with brown markings
Habitat:	Open sea

Jellyfish are most commonly seen around our shores in the autumn. At this time swarms of them may be swept on to the beach and left by the tide to rot along the strandline. Here they can be examined closely. However, care should be taken, for while the dainty pink common jellyfish and large barrel jellyfish are completely harmless to man, both the brown compass jellyfish and in particular the ice blue *Cyanea* can inflict a painful sting. If you become entangled in the stinging tentacles they are best dealt with by pouring alcohol or urine over them, which will prevent any further activity of the cells and stop further lesions from taking place; then seek medical attention.

All jellyfish are shaped rather like an umbrella. The upper bell gently propels the animal through the water by rhythmic pulsations. The edge of the bell may have a number of tentacles of varying length, together with tiny, light-sensitive sensors which tell the animal if it is going towards or away from bright light. Beneath the centre of the bell (the handle of the umbrella) are the mouth parts. These are divided into four or eight sections and hang down ready to catch any food that the jellyfish encounters during its passage through the open sea. Just like the closely related sea anemones, jellyfish use stinging cells on

JELLYFISH

their tentacles and mouth parts to paralyse their prey before eating it. They all eat either planktonic organisms or small animals such as young fish.

Most of the life of a jellyfish is spent in the open sea, but the reproductive process, which follows a similar pattern in all the different types, includes a period in which the animal is attached to rocks. Males liberate sperm into the sea and when it enters the female's brood chamber internal fertilization of her eggs takes place. This process is helped by the fact that jellyfish usually swarm together in large numbers and are therefore in close contact with one another.

The young develop in the often colourful, crescent-shaped, brood chambers until they are liberated through their parent's mouth in the autumn. They swim away, settle out on to part of the rocky coastline and develop into the next stage of their life cycle, when they look rather like sea anemones and are called scyphistoma. At this stage, the common jellyfish is only about 1.5cm in size but from it large numbers of tiny jellyfish bud off, each one dropping away like individual plates from a stack. Once free, they start to feed and grow into adults and it may be one to two years, depending on the species, before they are themselves ready to reproduce.

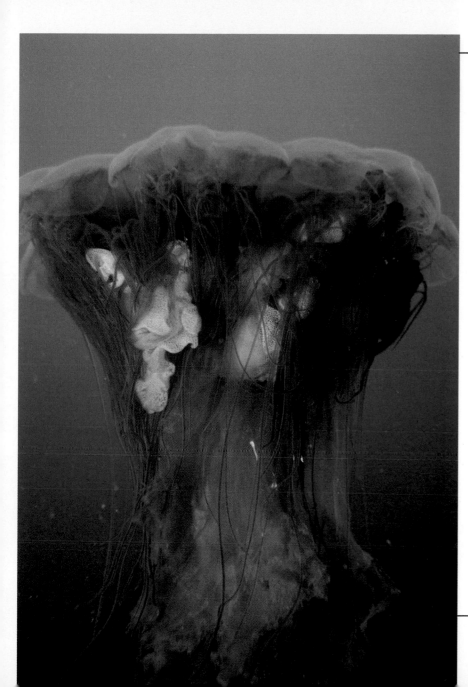

LION'S MANE JELLYFISH

Cyanea sp.	*Cyanea lamarckii*
Size:	Up to 15cm across bell
Colour:	Translucent blue/white
Habitat:	Open sea, southerly distribution

COMMON JELLYFISH

Aurelia aurita

Size:	Up to 25cm across bell
Colour:	Transparent with pink hue
Habitat:	Open sea and estuary

Facing page, top: Cyanea lamarckii *is one of the few stinging jellyfish. Lion's mane is a name also used for other 'stingers'.*

Facing page, bottom: Numerous tentacles trail behind the lion's mane jellyfish.

This page: From below, Cyanea *appear quite solid.*

Left: Sand is not a solid enough material for the beadlet to live on; a rock lies below it.

Facing page: Not only can beadlet anemones move around, they also fight!

Above: The blobs of jelly common on tidal rocks are actually beadlet anemones with tentacles withdrawn.

Right: The characteristic blue beads give Actinia equina *their common name.*

©Albina FELSKI

Self-taught artist **Albina Kosiec Felski** began painting in 1960. Like *The Circus*, her paintings are usually on four-by-four canvases and are brightly colored and extremely detailed. Though she has an easel, Felski prefers to paint with her canvas spread flat on a table. In this painting, she includes a myriad of circus acts, particularly those involving animals.

Albina Kosiec Felski, (Canadian, active in the United States, b. 1916)
The Circus (detail), 1971
Acrylic on canvas, 48 x 48 ¹/₄ in.
National Museum of American Art, Smithsonian Institution
Gift of Herbert Waide Hemphill, Jr. and
Museum Purchase made possible by Ralph Cross Johnson 1986.65.108

ISBN 1-56640-184-4

printed in the USA

Left: Sand is not a solid enough material for the beadlet to live on; a rock lies below it.

Facing page: Not only can beadlet anemones move around, they also fight!

Above: The blobs of jelly common on tidal rocks are actually beadlet anemones with tentacles withdrawn.

Right: The characteristic blue beads give Actinia equina *their common name.*

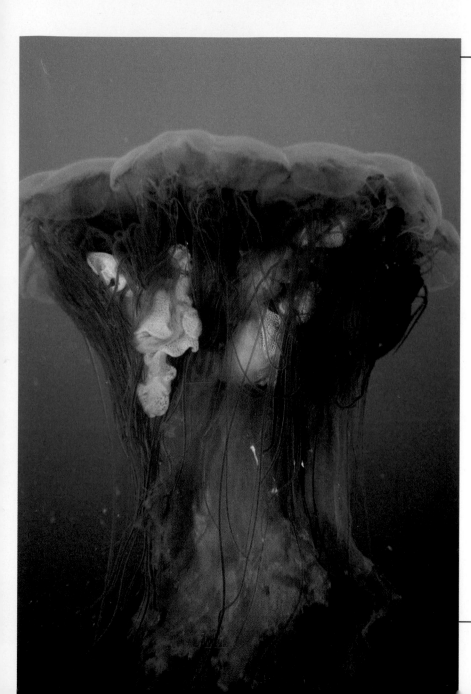

LION'S MANE JELLYFISH

Cyanea sp.	*Cyanea lamarckii*
Size:	Up to 15cm across bell
Colour:	Translucent blue/white
Habitat:	Open sea, southerly distribution

COMMON JELLYFISH

Aurelia aurita

Size:	Up to 25cm across bell
Colour:	Transparent with pink hue
Habitat:	Open sea and estuary

Facing page, top: Cyanea lamarckii *is one of the few stinging jellyfish. Lion's mane is a name also used for other 'stingers'.*

Facing page, bottom: Numerous tentacles trail behind the lion's mane jellyfish.

This page: From below, Cyanea *appear quite solid.*

BEADLET ANEMONE

Actinia equina

Size:	Up to 4cm tall
Colour:	Variable; red, brown and green are most common
Habitat:	Rocky shore and rockpools; middle shore and shallow sea to 8m

This, the best known of the sea anemones, lives on rocks and in pools at the edge of the sea. In the winter it seems to migrate down into deeper, warmer water to escape the cold of the beach. In summer it appears tolerant of heat and resists desiccation when the tide is out by withdrawing its tentacles into its body cavity. At this time it looks like a blob of jelly on the rock. Only when the tide floods back in, or if it is submerged in a rockpool, can you see the tentacles (there may be up to 192 of them) arranged in 5 or 6 rings around the mouth. Beneath the tentacles is a ring of 24–48 clear blue 'beads' – the number increases with age – from which the animal gets its name. Both tentacles and beads are covered in stinging cells. Those on the tentacles are used to paralyse prey, such as worms, prawns and small fish, which are then drawn into the body cavity and digested.

Beadlet anemones are either male or female and reproduce both sexually and asexually. Sexual reproduction results in planktonic larvae, and helps to disperse the species all around the coast. Asexual reproduction involves internal budding. The young developed in this way are brooded within the body cavity and then, when they are still only about 2mm tall and have only eight tentacles, they climb out of the mouth of their parent and make their way to a nearby piece of rock, where they can feed and grow.

Although most beadlet anemones are blood red in colour, they can also be green, grey, brown, pink, salmon, orange or strawberry. (Strawberry beadlets which are red with green spots are a separate species, *Actinia fragacea*, and do not brood young.) The foot (pedal disc) by which it attaches to rocks may be a different colour from the rest of the animal. In fact at times it has two colours which radiate out from the centre like the spokes of a wheel. Beadlets are surpisingly mobile animals, not only moving up and down the beach seasonally but often searching for a better position from which to find food. It is then that fierce territorial battles between individuals may take place. They lash out at one another with the batteries of stinging cells on the blue beads, until at last the loser gives up and moves away. In spite of the wounds they inflict on each other our skin is so thick that to us they only feel sticky.

BEADLET ANEMONE

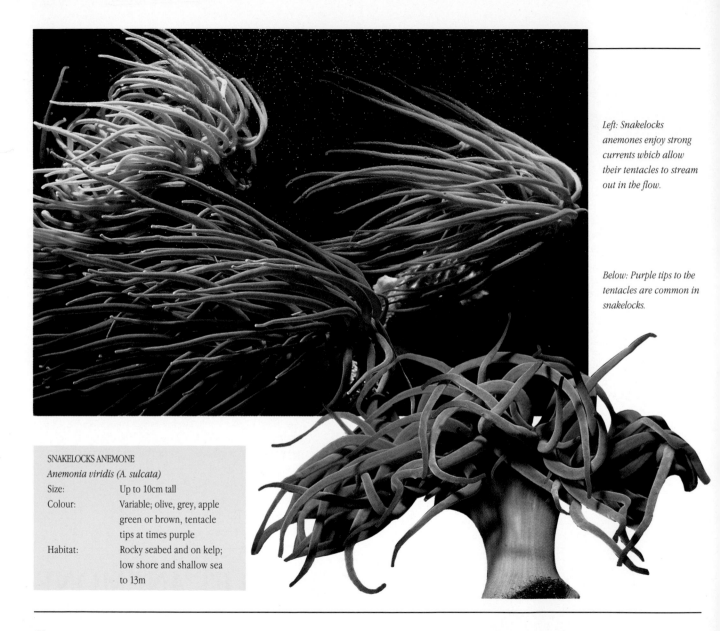

Left: Snakelocks anemones enjoy strong currents which allow their tentacles to stream out in the flow.

Below: Purple tips to the tentacles are common in snakelocks.

SNAKELOCKS ANEMONE
Anemonia viridis (A. sulcata)

Size:	Up to 10cm tall
Colour:	Variable; olive, grey, apple green or brown, tentacle tips at times purple
Habitat:	Rocky seabed and on kelp; low shore and shallow sea to 13m

Unable to withdraw its tentacles, the snakelocks looks bedraggled until the tide returns.

This anemone is unable to withdraw its 200 or so tentacles into the safety of the body cavity. As a result they constantly writhe about in the currents searching for food. This has given them the name 'snakelocks' after the mass of snakes on the gorgon's head.

These animals are predominantly olive brown, although they may at times be bright apple green in colour. Many have purple tips to their tentacles, which is where the greatest number of stinging cells are found. The body colour is enhanced by the presence of microscopic single-celled brown algae called *zooxanthellae* which actually inhabit the anemone's outer cells. Both plants and animals gain from their close relationship; the algae get protection and a rich supply of nutrients which they need to form organic compounds, while the sea anemones benefit from the removal of these chemicals.

To thrive the algae need plenty of light and snakelocks anemones are usually found in clear coastal water, especially on the western side of Britain and Ireland. They are among the few animals that do particularly well in tidal rapids and strong currents. Here they attach themselves either to the rocky seabed or to kelp fronds. At times, when conditions are favourable, colonies grow rapidly, reproducing both sexually and by simple division, so that hundreds of animals per square metre may carpet a rock shelf. Here they await their food which consists of small animals that live in the plankton and are brought in by the currents and tides.

Observation of these anemones has shown that they are territorial and like to be far enough away from one another so that the tentacles do not quite touch. If they get closer a battle will follow and it is usually the brown ones that hold their ground and the green that have to retreat.

SNAKELOCKS ANEMONE

Left: Each tentacle of the dahlia anemone has an array of stinging cells with which to ensnare prey.

Facing page: Dahlia anemones are colourful, large and plentiful. Their size and patterning make them easy to identify.

DAHLIA ANEMONE *Urticina felina*
Size:	Up to 12cm tall, occasionally larger in deep water
Colour:	Variable; radial bands of white, red, pink, green and blue
Habitat:	Shaded rocks and rockpools; mid- to low shore and shallow sea

Named after its colourful, flower-like appearence, the dahlia is the largest anemone found between the tides on the rocky shores around Britain. When contracted, the tentacles are withdrawn into the body cavity and the warty column has a covering of shell fragments and pieces of gravel. The dahlia tends to live in the shadows or under boulders but when seen in the sunlight, fully expanded, the colours can be simply stunning. The column can be green, pink or red, at times with grey markings. It also has conspicuous warts. On the closely related *Urticina eques* these can rarely be detected nor are there pieces of shell or gravel on the column. In contrast to the column, the somewhat transparent tentacles and the area around the mouth have bands of colour which may be shades of white, red, grey, cream, green or blue. The colours of the entire animal complement one another and are outstandingly beautiful.

The dahlia anemone copes well in exposed areas which are subject at times to considerable wave action. It is usually found firmly attached to rocks or boulders in low pools or on the sides of gullies and crevices. Once they have found a place they like they will stay there, often for years. The sexes are separate and fertilization of the gametes occurs in the sea after they have been released by the parents, in dribs and drabs throughout the year. From this planktonic larvae develop and will eventually settle out on a suitable rocky area. They are carnivorous and have about 160 short, stubby tentacles arranged around the central mouth with which they catch prawns, fish and other small animals.

DAHLIA ANEMONE

Left: Open plumose anemones have incredibly delicate tentacles, but are bland when closed.

Facing page: Plumose anemones can coat wrecks, giving them a ghostly white covering.

PLUMOSE ANEMONE
Metridium senile
Size: Up to 30cm tall
Colour: Variable; orange, pale cream,
             brown
Habitat: Rocky seabed and man-made
             structures; very low shore to
             100m

Plumose anemones are the largest found around the British Isles. They have smooth columns and thousands of slim tentacles encircling the centrally placed mouth. This gives them their feathery, plumose, appearance. It is these small tentacles that catch the planktonic creatures and suspended detritus on which the anemones feed.

Like many other species of anemone, plumose anemones come in a variety of colours, usually orange, brown or pale cream. They are mobile creatures and move slowly over the rock face. As they progress small bits of the column are left behind and from these torn edges new individuals arise. This means that it is quite usual to find a group of the same colour and genetic make-up. They live harmoniously together but will fight away any other anemone that comes too close.

Plumose anemones are usually found below low-tide level on rocks, in crevices or on man-made structures such as piers. They also colonize gas and oil rigs. Thousands of them live closely together forming a wide band below the intertidal band of mussels, and because they are smooth and slimy and have stinging tentacles, they form an effective barrier to starfish, which would normally climb up the rig legs to eat the mussels. If a rig is colonized too densely by vast numbers of heavy mussels there is a danger that installation checkpoints will be obscured or that there will be an increased drag on the structure. To avoid this it is not uncommon for divers to clear pathways through the anemones to allow starfish to climb up and eat the mussels.

PLUMOSE ANEMONE

Although often found in association with hermit crabs, the parasitic anemone can survive alone.

These anemones are not in fact parasites at all, rather they frequently enjoy a symbiotic relationship with hermit crabs, usually *Pagurus bernhardus*. Crabs and anemones can both live on their own, but there are benefits to both if they share their home. The crab gets protection from predators that are wary of the anemone's stinging tentacles. The anemone gets the advantage of a mobile home; it is moved by the crab to sandy areas and over other soft sediments that it could not enter on its own. It may also get some of the crab's leftovers.

The relationship is very close and when the crab moves to a larger shell it will encourage the anemone to go with it. To do this it drums the shell and the anemone with its claws, a process which apparently relaxes the hold the anemone has between its disc and the shell. Once detached, the crab will gently transfer the anemone to the new shell with its claws.

PARASITIC ANEMONE	
Calliactis parasitica	
Size:	Up to 8cm tall
Colour:	Creamy brown with vertical stripes, pale translucent tentacles
Habitat:	Rocky, muddy seabed, shells and whelk shells occupied by hermit crabs; shallow sea, 3-100m

PARASITIC ANEMONE

Left: Sagartia *anemones may occur in groups.*

Below: This individual S. elegans *var.* rosea *is one the most colourful anemones in temperate waters.*

Sagartia elegans	
Size:	Up to 4cm in diameter and 6cm high
Colour:	Variable; orange, brown, rose, with white or bicolour tentacles
Habitat:	Rocky seabed; midshore to 50m

*S*agartia elegans is a small anemone found widely around the coastline. Trying to identify it can be a little confusing simply because there are so many different colour forms. Five distinct variants have been described. These include variety *venusta* with an orange warty column and white tentacles, variety *miniata* in which both the tentacles and column are sandy brown and variety *rosea,* which is brilliantly coloured and almost tropical in appearance, with rose or magenta tentacles.

Sagartia are found in a similar habitat to *Actinothoë,* but rarely on the same rocks, and large numbers of them can be found covering rock crevices. Many of the animals that live close to one another may be genetically identical because, although sexual reproduction does take place, most new anemones grow from bits of the column that have been shed by an adult in the process called basal laceration.

These anemones use their 200 tentacles to catch their food from the plankton in the surrounding water. They themselves are quite well protected and will sting predators with special cells on the tentacles. They also have small white threads (acontia) which are shot out from inside the body cavity through tiny pores when they are greatly disturbed. These acontial filaments are covered with batteries of stinging cells which deter most aggressors. There are always some animals that are particularly determined and it is known that tompot blennies will shrug off these attacks and that *Sagartia* forms quite an important part of their diet.

SAGARTIA ELEGANS

Sagartia *in a brittle star bed.*

This delicate white anemone is striking, and quite common in the right habitat.

Individually *Actinothoë* are rather inconspicuous anemones and often pass unnoticed when lying alone beneath a rock or boulder on the low shore. There is, however, no mistaking them in their more usual position on submarine vertical cliff faces and rocky overhangs by exposed coasts. Here they carpet the rocks, the huge numbers dominating the surface. At times they are epizoic, which means that they live attached to other animals, most often Britain's largest sea squirt *Phallusia mammillata*, which grows to 12cm in height. The *Actinothoë* climb up to the top of the sea squirt, away from the silty, muddy seabed. In this position clean seawater will flow past, bringing with it planktonic food which can be caught by the 100 small white tentacles around the anemone's mouth.

ACTINOTHOË SPHYRODETA

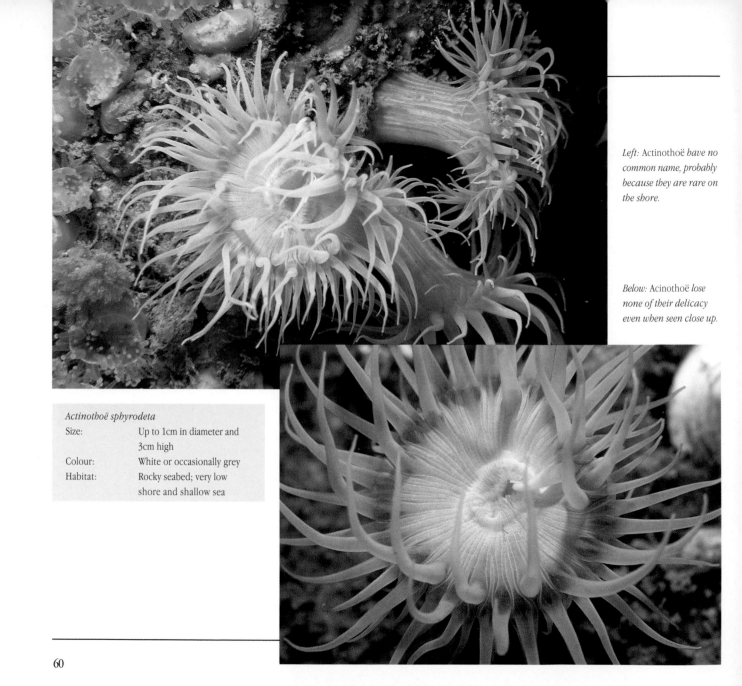

Left: Actinothoë *have no common name, probably because they are rare on the shore.*

Below: Acinothoë *lose none of their delicacy even when seen close up.*

Actinothoë sphyrodeta

Size:	Up to 1cm in diameter and 3cm high
Colour:	White or occasionally grey
Habitat:	Rocky seabed; very low shore and shallow sea

JEWEL ANEMONE	*Corynactis viridis*
Size:	Up to 1cm across
Colour:	Variable; purple, orange, pink and green with red/white tentacle tips
Habitat:	Rocky seabed; very low shore to 100m

J ewel anemones are a type of coral. They are found on rocks and boulders, in crevices and on the undersides of ledges. They are restricted to the south and west coasts of Britain and Ireland. At times they form large, bright, multicoloured colonies, individuals reproducing by simply dividing down their length (a process known as longitudinal fission). While they are true corals they do not have a hard skeleton, only a broad adhesive base to the column. They are unlike most other anemones in that each of the 100 or so tentacles ends in a small knob which stands out because it glistens, twinkles and is frequently a different colour from the rest of the animal.

Above: Different coloured jewel anemones may meet abruptly, producing a distinct border.

Left: Jewel anemones are very distinctive even when seen in a mass.

JEWEL ANEMONE

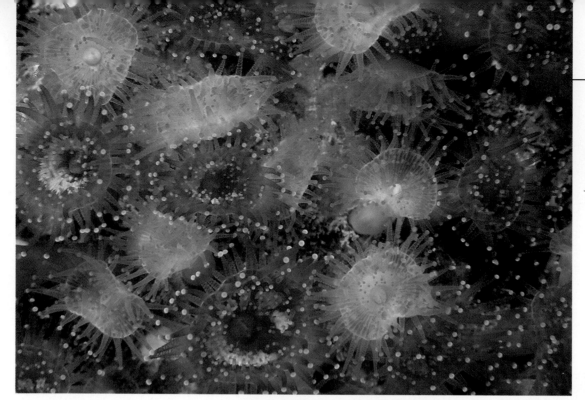

Left: Small, exquisite and beautifully coloured, jewel anemones are appropriately named.

Right: Where their colonization is not total, jewel anemones share the substrate with other colourful creatures.

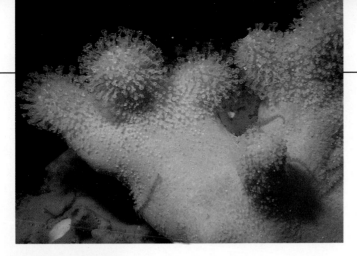

Right: Dead men's fingers can occur in either orange or white.

Below: Strong currents mean that food is constantly being washed past waiting creatures.

This branching, sponge-like organism is actually a colonial soft coral. It gets its name from its close resemblance to pale swollen fingers. The main 'body' of the colony is the flexible, tough, jelly-like skeleton. Within this soft structure lie embedded large irregularly shaped calcium carbonate spicules which help to protect the colony from damage.

Each member of the colony is called a polyp and looks like a tiny, 1cm high, anemone. Each individual is connected to its neighbour by a series of canals and each colony is usually all male or all female, although occasionally they are found to be hermaphroditic. While the skeleton may vary in colour, all the polyps are white and each has eight branched tentacles, which give them a rather feathery appearance. If there is any danger from predators or unfavourable sea conditions the polyps are able to close up and retract down into the skeleton.

Unlike an individual animal that continues to grow until reaching an optimal size, colonies of dead men's fingers and the number of polyps they contain vary throughout the year. They are largest during early summer, in May and June, when the seas around them are most full of the plankton on which they feed. As the food levels decline the colony stops feeding. During the non-feeding phase, which lasts from about July until November, the colony shrinks and eventually becomes only a quarter of its original size. At the same time the surface will have become leathery, rather tatty and covered in a variety of encrusting plants and animals. In December the whole of the outer layer is sloughed off and feeding starts again.

Dead men's fingers live attached to hard surfaces such as wrecks, boulders and rock faces in the shallow seas around the coastline. They are most often found in places where there is a lot of water movement and are strong enough to withstand currents of up to 1m per second. They become sexually mature in their second or third year and spawning takes place externally in midwinter. The larvae can remain in the plankton for long periods but it is more usual for them to metamorphose after just a few days.

DEAD MEN'S FINGERS

DEAD MEN'S FINGERS
Alcyonium digitatum
Size:	Up to 30cm high
Colour:	Variable; white, cream, orange, pink and grey
Habitat:	Rocky seabed; very low shore to 100m

Above: Extended polyps give the dead men's fingers a haloed appearance.

Right: In close-up the polyps assume a more distinct identity.

Facing page: These brittle stars have found a convenient way to access the food-providing current!

Mussels attach
themselves to solid
surfaces, and are in turn
used as such by
barnacles.

MUSSEL	*Mytilus edulis*
Size:	Up to 14cm
Colour:	Shell blue-black or brown outside, pearly inside
Habitat:	Rocks and rocky shore, at times in dense beds; mid-shore and shallow sea

Byssus threads act as anchoring 'guy-ropes' for mussels.

Mussels are very common all around our shores. They have a shell of two halves (valves) which hinge at the back and are held tightly together by strong muscles. Although a close relative of both the herbivorous limpet and the predatory, carnivorous octopus, mussels gather their food by filtering seawater through the gills at a rate of up to 2 litres per hour. The gills are able to grade, sieve out and move plankton to the mouth, in addition to their more normal role in breathing.

In sheltered areas such as sea lochs, mussels grow rapidly and may reach 10–14cm in length in about two years. This is in contrast to mussels found on exposed rocky shores to the west of the country, where the waves are so powerful that growth is severely restricted and the animals may only be 2 or 3cm after five or more years.

Mussels always tend to settle close to one another and grow in clumps, but on very exposed rocks they are packed tightly together, each hanging on to its neighbours and the rock below by hundreds of fine threads, called byssus. The byssus threads act like guy ropes around a tent. They allow the animal to bounce a little in the waves but prevent the force of the waves from sweeping it away. If, during a time of less wave impact the mussel decides to relocate itself, it is able to discard the byssus threads and crawl, using its single muscular foot, to a new site. Sometimes, it will even make and break threads as it walks to prevent the possibility of being carried away by the sea. When it decides on its new location it will rapidly establish itself by making more attachment threads. The same process occurs on a mussel bed. Here, thousands of animals live in crowded conditions, producing a hugh volume of waste material that settles down and would eventually smother the lowest animals. To prevent this from happening, mussels continually crawl from the bottom of a clump to the top, where there is less mud and more food.

Any hard surface around the coast is a handy site for marine organisms to settle and grow. Mussel shells are no exception to this. The animals that most often take advantage of mussels as a settlement site are barnacles. Like the mussels, barnacles also have a mobile larva, but once settled they change into non mobile adults. At first glance it would seem that they run the risk of squashing each other as they grow. In fact this is avoided by barnacles of the same species because they always choose to settle a few millimetres from one another.

MUSSEL

This page: Scallops have distinctively shaped shells which have often been used as an artistic motif.

Facing page: Small, gleaming 'eyes' allow queen scallops to detect potential predators.

QUEEN SCALLOP	*Chlamys opercularis*
Size:	Shell up to 9cm long
Colour:	Variable; purple, brown, red or yellow
Habitat:	Sandy, gravel seabed; very low shore to 200m

S callops have been a source of fascination and inspiration for centuries. Originally collected for their sweet, delicious meat they soon became a symbol widely used by artists both in the classical world and in the Renaissance. Later they became a popular motif on heraldic shields and from as early as the twelfth century the wearing of a scallop shell became the mark of an individual who was making a pilgrimage to the tomb of St James at Compostela in Spain.

Queen scallops are often called queenies and do not grow very big, only reaching about 9cm in diameter. Their shells are made of two valves. The lower one is the deepest and has the body sitting in it whereas the upper one is flat and acts as a protective lid. Queenies are gregarious creatures and are found in their thousands in regions where the seabed is sandy or muddy. A close look will show that at rest the valves gape a little. This enables them to breathe by pumping water over the gills while watching for danger with their hundred or so small eyes. These little eyes, which lie along the edge of the shell between the tentacles are simple, pigment-spot ocelli. Although they are unable to see an image clearly in the way we can, they are able to detect changes in light intensity. If a shadow passes over them it could

mean an approaching predator and they hurry to escape.

Scallops move by clapping the shell valves together like castanets. This pushes water out from the shell and thrusts the animal through the water. The effect of this is that each scallop looks as if it is bouncing backwards up and down through the sea. They have a remarkable degree of control over the direction in which they go and when watched closely can be seen to swim normally, or to twist, or to undertake 'escape' movements. The ability to move about freely means that scallops can migrate to different feeding areas at different times of the year. However, the main advantage is that they can escape predators. Starfish

are the main threat because they are strong enough to pull the shell open with their powerful tube feet.

The mature scallops breed between January and June, releasing their eggs and sperm into the water around the time of the full moon. This synchronization helps to ensure that as many eggs as possible are fertilized. When the small larvae first settle on a firm substrate on the seabed they crawl about using their feet. After a while they choose a suitable site and attach themselves by means of byssus threads, just as mussels do. However, unlike many other species of mollusc, which remain in one place for life, the queenies soon become detached and spend the rest of their lives free on the seabed.

QUEEN SCALLOP

LESSER OCTOPUS *Eledone cirrhosa*

Size:	Up to 50cm long; one row of suckers on the tentacles
Colour:	Orange/rust above, pale below
Habitat:	Rocks and boulders; low shore and shallow sea

The dappled skin of the octopus serves as a camouflage and can undergo rapid colour changes.

Right: Octopuses have good eyesight and can see almost as well as humans.

Below: Body colour changes are accomplished by the muscular contraction of individual pigment cells.

Octopuses are some of the most intelligent and most intriguing of all sea creatures. They are closely related to both cuttlefish, which have an internal shell often used to feed budgerigars, and squid, which have a thin glass-like skeletal plate. Octopuses differ in that they have no skeleton at all. The body is rather bag-like and extremely flexible, which means that they can squeeze through tiny nooks and crannies between the rocks and boulders among which they make their home.

Of the two types of octopus found in British waters, it is the widely distributed lesser octopus that is most likely to be found on the shore, particularly in summer. The common octopus is usually restricted to the English Channel where it can be found in plague proportions after a mild winter. Both prefer cool, deep water and struggle to survive in summer if caught in a pool in which the temperature rises much above 18°C. Like squid and cuttlefish, octopuses are not only camouflaged by their dappled skin, but are also able to undergo rapid changes in body colour. This occurs by muscular contraction of individual pigment cells. When relaxed, the cell is round and little colour

OCTOPUS

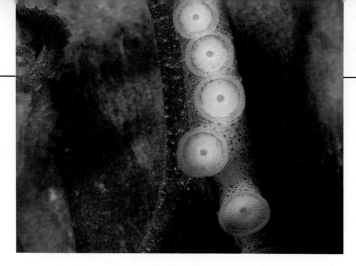

The octopus creeps across the sea floor on eight sucker-covered tentacles.

is seen; when contracted, it is pulled out into a round disc of bright colour. It is thought that the changes are caused by the nervous system and certainly octopuses adjust colour to their mood. When alarmed they present an elaborate defensive colour change, in which rapid variations from light to dark occur and dark spots appear around the eyes.

There are two ways in which they move about. Normally they creep over the sea floor on their four pairs of sucker-covered tentacles. However, if they want to move faster they change tactics and take in a large breath of water which they shoot out of their

siphon under such pressure that the whole animal jets rapidly backwards. At times, when escaping from a threat or a predator, they will also liberate ink from a sac so that they effectively disappear behind a blue-black cloud.

Although they are possibly a little short-sighted, octopuses have good eye sight and can see almost as well as we can. This, together with their intelligence means that they are very effective predators. They have no skeleton but their jaws are like a hard, horny parrot's beak and are more than able to bite open shells. If you are ever privileged to handle one of

these animals, do beware, as they can give a nasty nip!

Once the prey is spotted the octopus rises up and leaps on to it, clasping it firmly with all eight tentacles. In fact, their favourite food is crabs and they soon learn to approach them from behind so as to avoid their claws.

Octopuses become sexually mature at about two years, and it is interesting to note the care given to the eggs compared with their relatives the squid and cuttlefish. The ocean-going squid simply liberates large clusters of club-shaped eggs and, once they are fertilized, leaves them to develop on the seabed. Cuttlefish lay a hundred or more single eggs, each resembling a large black grape. The female attaches them, often in groups, by a small stalk to a piece of weed or rock. In a similar way octopuses will also lay clusters of eggs, again resembling a bunch of grapes, usually in a rocky crevice. In both species the female will remain close by her eggs and tend them carefully as they develop. After laying her eggs the octopus will stop feeding and will expend all her energy on guarding them from the many predators eager to snatch an easy meal. She will remove the debris or sediment that continually collects on them and will constantly blow fresh seawater over them to improve aeration. Eventually, when the young have hatched, the female, at least of the common octopus, is so weakened by brooding her eggs that she dies.

Their intelligence makes octopuses very effective predators.

cools with the onset of autumn, growth slows and they move offshore to deeper water with a more stable overwintering temperature.

Prawns move in a number of ways. They walk forward using the last three pairs of legs; they dart backwards by jet propulsion, facilitated by a sudden, rapid flexing of the tail; or they swim through the sea by beating the five pairs of paddle-like appendages (swimmerets) beneath the tail.

The swimmerets also have another function. In summer, after moulting, the female's ovaries are ripe and she is receptive to the male. He will fertilize the 2,000–4,000 eggs which she then cements to the swimmerets. Here they are brooded for up to four months. They then undergo a short planktonic stage before they settle on to the rocky seabed in July and August.

During the day prawns rest quietly beneath weed or among rocks, where they are well camouflaged. It is at night, after the rock pools have cooled from the heat of the summer sun, that they become active. They emerge to feed off bits of seaweed and other materials, which they delicately tear to shreds with the tiny nippers on their first pair of walking legs.

Above: Prawns have almost transparent bodies, both eyes sometimes being visible from one side.

Below: The shed carapace of a prawn is exquisitely delicate. This one survived as it hung from kelp.

Prawns are most common to the west and south of Britain, and are usually to be found on the rocky shore in summer. They may reach 5cm in length and are easy to tell from shrimps because they have a rostrum (sword-like spine) between the eyes. However, more detailed observation of the shape of the shell and the size or number of spines is needed to identify individual species. Because they are almost transparent, it is easy to miss them as they hide amongst seaweed in gullies and shallow pools. They can live for 3–4 years and grow most rapidly in the warm summer sea when they moult their shell every two weeks. As the sea

COMMON PRAWN

Above: A prawn's pincers are relatively much smaller and less powerful than those of many of their relations.

Above: In spring, young prawns can be seen in their thousands in shallow rockpools.

Right: Crevices are ideal homes for prawns.

COMMON PRAWN	*Palaemon serratus*
Size:	Up to 7cm
Colour:	Almost transparent, slight vertical brown banding.
Habitat:	Rockpools and rocky shore; low shore and shallow sea to 40m

Right: Crawfish are not as aggressive as lobsters, but will fight if required.

Below: Crawfish are also called 'spiny lobsters'.

Although closely related to lobsters, crawfish are easy to distinguish from them in a number of ways. They are orange not blue and have beautifully sculptured shells with large spines. In addition, none of the five pairs of walking legs has pincers (except in females where they are found on the fifth pair). Occasionally they can be found in shallow water or among boulders on the shore but crawfish usually live in deeper water than lobsters. Both reside in similar rocky burrows, to which they return at night. Crawfish emerge under cover of darkness to scavenge and it has been found that they are able to find their way home, even when blindfolded! How they do this is not really known, but it is most likely to be by a combination of taste, sound and directional current flow within the sea. Their sense of smell is certainly very well developed and it is known that they detect their food by means of the chemical receptors on their legs.

Unlike the solitary lobster, crawfish usually live together in large groups. They each have their own burrow or crevice in the rock face and are territorial, fighting any other crawfish that enters their home area. When feeding, twenty or more may march across the seabed in search of molluscs, crustaceans and other types of prey. At night in the deep sea, eyesight is of little use and the animals keep in contact with each other by sound. They emit a grating noise which they make by drawing their bristle-covered antennae over a ridge which protrudes beneath the eye.

Apart from their spiny, hard shells, the antennae are the main form of protection. They can be used as sharp-edged whips that can lacerate the soft flesh of a predator, although some fish, such as the trigger fish, are alert and will simply nip them off.

CRAWFISH	*Palinurus vulgaris*
Size:	Up to 50cm in length with antennae of a similar length
Colour:	Orange/brown
Habitat:	Rocky sea shore to the south and west of Britain; shallow sea to 70m

CRAWFISH

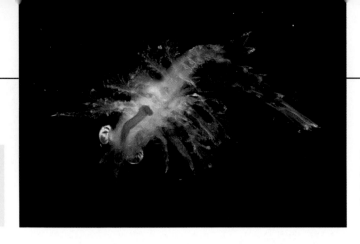

LOBSTER	*Homarus gammarus*
Size:	Up to 45cm
Colour:	Blue/black
Habitat:	Rocky or boulder-strewn seabed

Left: This larval lobster is partway through shedding an outgrown carapace.

Blue carapaces and black eyes are characteristic of young lobsters.

T his beautiful, slow-growing crustacean is perhaps best known as a delicacy on the rich man's table. Although familiar to most they are rarely seen because of their marvellous camouflage coloration and the fact that although they do sometimes move about by day they are predominantly nocturnal, particularly the juveniles. Lobsters are solitary animals and only meet when about to mate. At other times they establish a territory and keep to it, carefully avoiding any confrontation with another lobster. They tend to remain in their burrows, only coming out under cover of darkness to forage for food. Lobsters are omnivorous scavengers and their diet includes dead and live material ranging from worms to fish. They have tiny mouths and use the larger, heavier front leg or claw to crush the food and then cut it with the lighter straight claw. They then shread it further with the second and third pairs of legs which also have articulated pincers, and finally the mouth parts, with their teeth-like edges, reduce the food to edible size.

The whole of the lobster is covered by a hard outer shell. Over the stalked, compound eyes the shell is almost like glass, which enables the animals to make

Right: While only 2 or 3 cm long, lobsters already display aggressive instincts.

Below: Juvenile lobsters are generally nocturnal, and therefore seldom seen.

out the rough shape, form and size of an object. Over the rest of the body it is jointed and heavily calcified, and protects the animal, as a suit of armour protected a knight. Like other crustaceans, the lobster can only grow by shedding its shell and forming a new one. This process tends to occur after a period of fasting, in a sheltered burrow and under the cover of darkness. The old shell breaks across the back and the soft animal pulls itself out and swells up until it is about 1cm longer than before. Slowly the flexible, soft

new shell will become calcified so that after a week or two the new larger shell is as hard as the old and is able once again to protect the animal from predators. Ageing a wild lobster is very difficult because growth is affected by water temperature, local conditions and the availability of food. Growth by moulting continues throughout its life. However, as it becomes older the rate slows from 3–4 times during the first couple of years to once a year until it is about 15 years old. After this, and it is thought that in secluded places with

LOBSTER

little fishing pressure lobsters may live for 60 or more years, moulting may only happen every 3–4 years.

Females become sexually mature when they have attained a carapace length (the carapace is the first section of shell) of around 70mm at an age af about 5–6 years. It is to allow all lobsters to breed at least once that the British legal minimum catch size has been set at a carapace length of 85mm. As with other decapods, it is only after recent moulting that a mature female is receptive to a male. At this time she produces a pheromone (a chemical signal) that attracts a male to her. They mate belly to belly and he deposits a packet of sperm that she will retain for use after her shell has hardened and she is liberating the eggs. Her tail is a little wider than that of the male and the eggs, up to 20,000 in a large female, are glued on to the swimming appendages. This all happens in late summer and she then carries the eggs carefully tucked under her tail and maintains them by gently wafting seawater over them which aerates them. The following spring, in May or June, the eggs will have matured and they will hatch just after sunset. The young larvae, only about 2mm in length, are attracted to light and swim up to the sea surface towards the moonlight. Their mother helps them by standing with her tail raised and by beating the swimmerets so that the larvae are thrust out into the sea. This is contrary to her normal behaviour, as she usually eats any larvae that come within reach.

Life for the young lobster is perilous and few of the thousands of larvae will survive their 4–5 weeks in the plankton. Those that do are voracious feeders and grow rapidly, moulting four times until, at about 1cm, they look like small lobsters with large claws and are able to make their way down to the seabed. Here they excavate little burrows for themselves and establish a territory.

A close-up view of the lobster's characteristically black eye.

Right: This lobster has only one claw and both antennae are broken, probably due to fighting.

Aggressive behaviour is quite normal among lobsters.

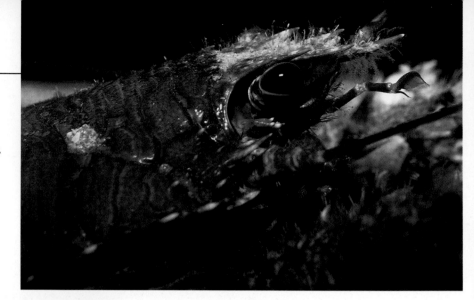

Right: The squat lobster,
Galathea strigosa *is colourful with sharp spines on its carapace.*

Below: The duller
G. squamifera *inhabits rockpools.*

SQUAT LOBSTER	*Galathea squamifera*
Size:	Up to 5cm in length
Colour:	Olive green/brown
Habitat:	Rocky shore and seabed; low shore to 80m

SQUAT LOBSTER	*Galathea strigosa*
Size:	Up to 12cm
Colour:	Orange/red with blue bands
Habitat:	Rocky shore and seabed; low shore to 35m

*Squat lobsters cling
upside down, hanging
under ledges.*

Squat lobsters or squatties may look as if they are a cross between a crab and a lobster, but are actually more closely related to hermit crabs. These small decapod crustaceans have flattened bodies, which are ideal in helping them to hide in crevices among rocks, stones and kelp holdfasts. They usually move by crawling on four pairs of legs but carry their tails tucked under their bodies and can dart rapidly backwards by giving it a quick flap.

There are five species of squat lobster found regularly around the British coast. Of these it is the small olive *Galathea squamifera* and the beautiful, brightly coloured *G. strigosa* that are most commonly seen, especially in spring and summer. Both have a sharply pointed spiny rostrum between the eyes, and front legs that are one and a half times as long as the body. These front legs, ending as they do in sharp hairy pincers, indicate accurately the way in which the animal feeds. It picks away at anything edible it encounters, and will hunt over a considerable distance to find food. At times, this even leads it into baited prawn pots and considerable numbers of the larger *G. strigosa* are landed with the commercial prawn catch.

SQUAT LOBSTER

COMMON HERMIT CRAB
Pagurus bernhardus
Size:	Up to 3cm intertidally, 10cm in deeper water
Colour:	Orange
Habitat:	Rocky shore and sandy seabed; lower shore to 140m

In close-up, the hermit crab has a fascinating, complex 'face'.

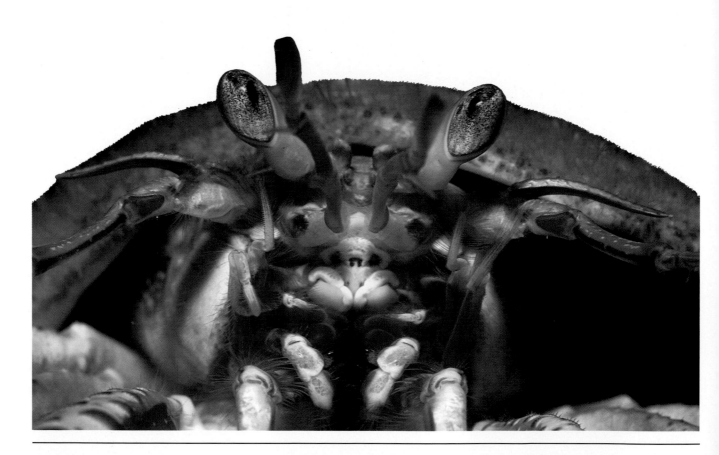

Commonly found on the south and west coasts of Britain, hermit crabs are easy to recognize because they always live in empty gastropod shells. They grow rapidly and in the process moult their carapace just like other crabs. This means that they have to change their borrowed shells regularly because they quickly outgrow the old ones. Usually they start life by inhabiting winkle shells and move up to the larger whelk shells as they grow. To move from one shell to another without falling victim to a hungry, passing fish, the crabs use their largest pair of legs and exploratory antennae to flip rapidly from one shell to another.

Hermit crabs are very particular about the shells they choose and spend a lot of time trying out alternatives. They test them carefully for both weight and size so that they get the best possible fit within the spiral shell for their coiled, soft, delicate, unprotected tail.

As well as using the borrowed shell as a home, hermit crabs like to camouflage it. To do this they actively encourage an assortment of animals to share it. These can include orange or yellow sponges, small colonial cousins of sea anemones called hydroids, which give

In common with other crabs, the hermit has a multi-faceted eye.

the shell a woolly appearance, sea anemones (especially the large colourful *Calliactis parasitica*), and the segmented ragworm *Nereis fucata*, which emerges from time to time to eat scraps of food. All these creatures provide help in some way: confusing predators, maintaining shell cleanliness or protecting the crab. Close observation may often reveal a yellow lump under the body. This is the reproductive stage of a parasitic barnacle called *Peltogaster paguri* which may weaken but rarely kills its host and is seldom seen unless the crab is actually in the process of changing shells.

Hermit crabs live on the seabed in areas that are soft and sandy and where the water currents are not so strong as to disturb them as they scuttle about. On occasion they can also be seen on top of submerged

HERMIT CRAB

boulders. This is most usual in the evening just as they are emerging to feed for the night. Although it must make them somewhat more vulnerable to predators it probably gives them a chance to see their surroundings or to 'taste' the water to determine the best direction in which to go hunting.

Like most crabs, hermit crabs are scavengers and eagerly rip pieces of food apart with their pincers. They grow quickly and can reach sexual maturity after about one year. At this age they only carry one brood a year but as they grow longer and larger they may have several. Once laid, the eggs are carried for two months on the reduced, non-functional swimmerets, which remain on only one side of the body. After hatching, the larvae swim and feed in the plankton for about two months before they settle down to life in the first of the many empty shells which are to become their home.

Above: Hermit crabs borrow shells with which to protect their soft bodies.

Right: An encrusted shell serves to further increase the hermit crab's ability to hide.

Spines and stalked green eyes give the spiny spider crab a fascinating appearance.

With small bodies and long legs, these crabs superficially resemble spiders. They are also unusual in that whereas most crabs can only walk sideways, spider crabs can also move diagonally. Again, unlike most other crabs which protect themselves by burrowing down into sand, spider crabs camouflage themselves by attaching an assortment of animals and plants to their shells. Tiny long-legged spider crabs are usually covered in sponges, red seaweeds and the occasional young anemone, whereas bigger spider crabs may also have large colonies of hydroids on them and at times barnacles may settle on spiny spider crabs. All have small spines on to which this decoration is placed and as soon as the animals have moulted, they carefully transfer many of the organisms from the old empty carapace into their mouths, where they are coated with a sticky substance, and then on to the new shells.

The smaller types of spider crabs are fairly sluggish and usually live among rocks and seaweed, where they pick away at their food with the pincers at the end of their long, delicate legs. They are omnivorous, feeding on a variety of seaweeds and animals. Their limbs are easily damaged and it is not at all unusual to see animals from which one or more legs have been lost. Not until the next moult will these legs be regenerated. The larger spiny spider crab is far more robust and its pincers are used both for defence and for catching and holding prey. They are common off

LONG-LEGGED SPIDER CRAB

Macropodia rostrata

Size:	Up to 1.5cm
Colour:	Yellow ochre
Habitat:	Rocky seabed; low shore down to 120m

SPIDER CRAB

Left: The spiny spider crab is the largest of the spider crabs and the easiest to identify.

Left: Macropodia *sp. have very long pincers giving them an unbalanced look at times.*

Above: It can be difficult to distinguish crab from covering!

SPINY SPIDER CRAB
Maja squinado
Size: Up to 18cm
Colour: Rusty orange
Habitat: Sandy seabed and rocky
             shore; low shore down to
             75m

Delicate limbs are easily damaged by aggressive behaviour.

Top left: Devoid of a weed covering, Hyas araneus *strikes up a defensive posture.*

Right: Some movie creations are hardly more impressive than this close-up of Hyas araneus!

Bottom left: This spider crab is well camouflaged among the weeds.

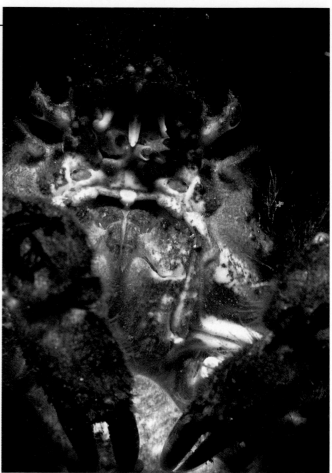

SPIDER CRAB	*Hyas araneus*
Size:	Up to 7cm
Colour:	Dull brown
Habitat:	Low shore and shallow sea to 80m

SCORPION SPIDER CRAB
Inachus dorsettensis

Size:	Up to 3cm
Colour:	Red to brown
Habitat:	Muddy, sandy and rocky seashore and seabed; very low shore to 180m

south-western coasts, where they patrol the sandy seabed for their food, which consists of a mixture of small animals and seaweeds. Although they look fairly cumbersome, once they have caught the scent of their food they march rapidly and directly up to it.

Spiny spider crabs have a fascinating courtship ritual. During July and August, in shallow water, up to 100 crabs meet together. It is quite common to find that the older males stand guard on the outside, protecting the rest of the group from predators. Inside this encirclement the younger males then mate with newly moulted females. After this it is another six months before the female will lay her eggs, which she then carries beneath her tail flap for nine months. Finally the eggs hatch into larvae which live in the plankton for a while before descending to the seabed and taking up an adult life.

Above: While a covering of sponge can act as camouflage, it may sometimes have the opposite effect.

Right: The scorpion spider crab is omnivorous, feeding on a variety of plant and animal life.

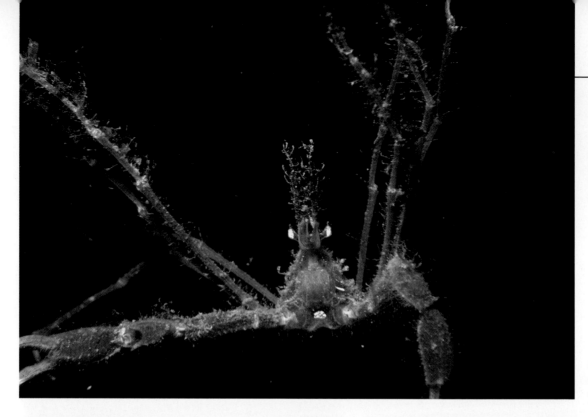

Left: Being unable to swim does not mean a lack of confidence in water!

Below left and right: Sponge-encrusted scorpion spider crabs.

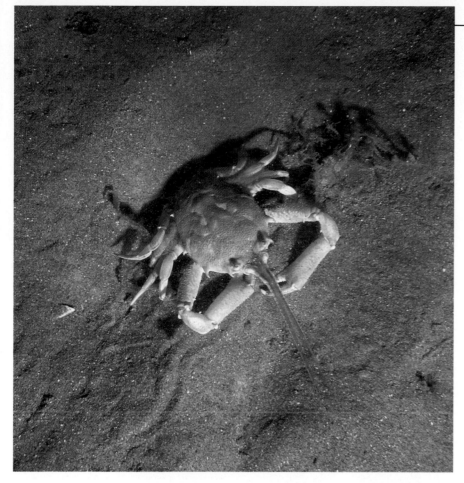

Masked crabs live in sandy habitats.

While most crabs occasionally bury themselves, masked crabs spend almost their entire life beneath the seabed. They lie buried during the day away from the watchful eyes of predators and then emerge at night to forage for food. They are only found in clean, well-graded sand, and are specially adapted to this environment. Like many marine animals, masked crabs use their gills to breathe. To prevent them from clogging with sand a large pair of antennae, which are as long as the body, project through the sand to the surface. These antennae have a series of hairs which inter-knit so that they form a breathing tube which acts rather like a snorkle and enables sediment-free seawater to circulate to the gills.

When placed on sand masked crabs do not scurry away sideways, as other crabs would, but rather start to dig down backwards. To excavate the sand they use the rear four pairs of legs, each of which has a long claw, to shovel the sand out of the way.

MASKED CRAB	*Corystes cassivelaunus*
Size:	Up to 4cm
Colour:	Pale yellow to brown
Habitat:	Sandy seabed; low shore and shallow sea to 60m

MASKED CRAB

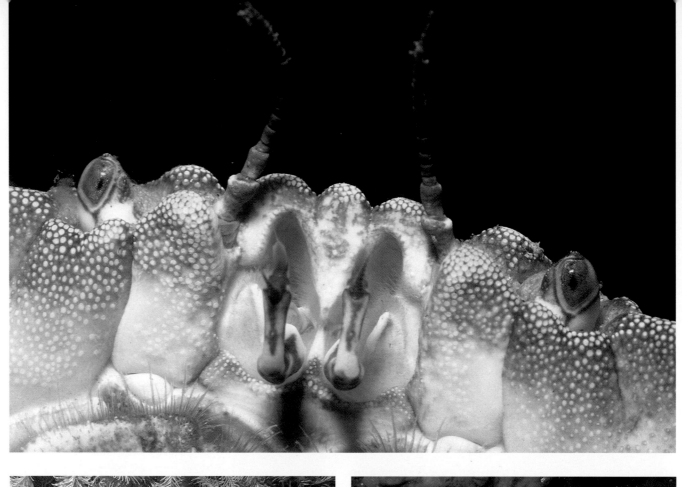

Left: Shallow boulders covered in seaweed provide homes for many creatures including Cancer pagurus.
Facing page, top: Edible crabs have small green eyes.
Bottom left: This edible crab is digging into the muddy seabed in order to hide.
Bottom right: Out in the open, the distinctive black claws of the edible crab are visible.

This is the largest crab in Atlantic European waters and is highly sought after because of its commercial value. At times the wide, oval shell can be as much as 30cm across and the animal can weigh 5kg. The shell has a distinctive crimped edge, reminiscent of an old-fashioned piecrust. The first pair of legs are muscular pincers, black tipped and so strong that they can easily crack open mussel and oyster shells so that the animal inside can be devoured. Despite the strength of the pincers and the fact that they can give a painful nip, when an edible crab is picked up it is usually quite docile. What happens is that they tuck their legs under themselves, effectively rolling into a ball and 'playing possum'. Once put back down on the beach they will stay still for a few minutes before burrowing into the sand or scurrying away to safety.

EDIBLE CRAB	*Cancer pagurus*
Size:	Up to 30cm across shell
Colour:	Orange/brown
Habitat:	Rocky sea shore; low shore to 90m

EDIBLE CRAB

The carapace of Cancer pagurus *looks like an old-fashioned pie-crust.*

Like all crabs, the edible crab has quite a complex life cycle with three distinct body forms throughout its life. Mature adults mate in late summer after the females have moulted. The eggs at this stage are neither laid nor fertilized, as each female carries the packet of sperm she has received from the male until she is ready to use it in December. The eggs, up to three million in a large crab, are then carried beneath her tail, where they gradually mature until they are ready to hatch between April and August. This is the ideal time for the larvae to develop in the sea because there is a lot of plankton to feed on.

The first stage in the development of the crab is the zoea larva, a spiny, large-eyed transparent creature which looks totally unlike its parent. The second stage is called the megalops larva. This looks a little more crab-like and has all ten legs, a larger body and huge eyes. After about 30 days in the plankton, the larva moults its shell to take on the characteristics of an adult crab, together with its seabed-living lifestyle. Young edible crabs live among kelp holdfasts and under boulders on the rocky shore. Here they scavenge for anything edible. They moult frequently and grow quickly, living for up to 20 years. During the winter they can be quite difficult to find on the shore. The reason for this is that they take advantage of the fact that in summer the coastal water is warmer than the open sea but that in winter the reverse is true. As a result they tend to migrate from the shore to deeper, warmer water in the winter. The migration is also tied into the breeding cycle and for example in the English Channel, crabs move west to sites off the Devonshire coast. Here the female finds a substrate which is not too coarse and which will not damage the eggs as she buries herself. It is also an ideal position from which the larvae will drift east with the current to restock the channel.

Larger, older animals tend to move further and further from the shore and it is these crabs that are fished commercially, usually with pots baited with pieces of fish. A minimum catch size, measured across the shell, is enforced around Britain in the hope that it will prevent overfishing and a reduction in the population.

Top: Although not a 'swimming' crab, the shore crab can have a go.

Bottom: Coloration is ideal in the weedy, rocky environment favoured by the shore crab.

T he shore crab is the one most commonly found and most easily recognized during a visit to the seaside. Young animals, only a centimetre or so in size, hide from the sunshine beneath rocks and in rockpool crevices. Larger crabs are found lower on the shore beneath seaweed or buried in sandy mud. Shore crabs can withstand a wide range of conditions, including rapid changes in temperature and salinity. As a result they are among the hardiest of shore animals, living in a wide variety of places, from small rocky bays to muddy estuaries and even polluted docks.

Shore crabs epitomize decapod crustaceans. They are ferocious creatures despite their small size and it is with good reason that the French give them the name *le crabe enragé*. They can give a painful nip with their pincers and are best held beyond nipping reach, across the back of the rough-edged shell. Underneath

SHORE CRAB

SHORE CRAB	*Carcinus maenas*
Size:	Up to 8cm across shell
Colour:	Olive green or brown
Habitat:	Rocky shore and sandy
	shore; low water to 60m

The shore crab is probably the best known since it is easily found in rockpools.

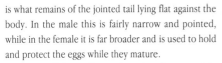
Careful study of the shore crab shows that its carapace acts as a home for various encrusting animals.

is what remains of the jointed tail lying flat against the body. In the male this is fairly narrow and pointed, while in the female it is far broader and is used to hold and protect the eggs while they mature.

However, if the crab is infested with a parasitic barnacle called *Sacculina,* the parasite forms a rounded mass below the tail flap and makes males take on the wide-tailed female characteristic. This parasite may remain in the crab for 3–4 years and during that time it takes up so much of its host's resources that the crab is unable to moult and grow. Only when the parasite eventually dies and drops off will the normal pattern of growth resume.

Normally shore crabs only live for about four years, but they mature early, mating for the first time when only a year old. As with most crustacea, the female is only receptive to the male when she has moulted. He makes sure of his success by hunting for a female which is about to moult and clasping her. They remain together in this paired state for several days, the female carried beneath the male, until moulting and mating have been successfully completed. The eggs, up to 185,000 of them, are then laid and brooded. They hatch, go through a three-month planktonic phase (during which they disperse along the coast), and finally settle on the shore.

Although they are quite easy to find at the seaside during the day, it is at night that shore crabs are most active. They have an acute sense of smell and patrol the seabed to hunt and scavenge for food, eating most of what they encounter, including worms, molluscs and dead animals. Once they find something suitable, they rip it apart with their pincers and then grip and manipulate it with the external jaws that surround their mouths. By nature they are opportunists but studies have shown that when they are foraging amongst a mussel bed they display an interesting tendency. They choose to crack open mussels of a similar size, so that they get the largest amount of food for the least amount of effort.

Left: This swimming crab is apparently digging a hiding place in the sand.

While possessing similar coloration to the edible crab, the swimming crab is actually very different.

SWIMMING CRAB	*Macropipus depurator*
Size:	Up to 6cm across shell
Colour:	Brick red to pale rust
Habitat:	Sandy seabed; 5 – 20m below low-tide level

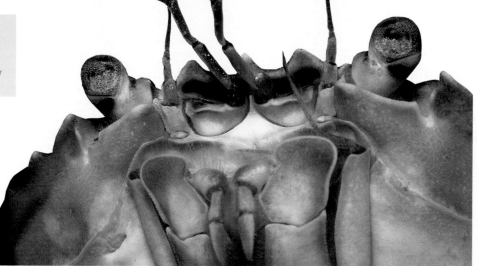

*Like its velvet cousin,
the swimming crab is
always ready for a fight.*

S wimming crabs are found on sandy seabeds, often dug into the sand itself. They have a smooth shell and brown eyes, in between which there are three sharp spines.

Swimming crabs have two methods by which they can move around: they can walk, or as their name suggests swim. They are able to do this because the last two pairs of legs are broad, flat and paddle-like. By beating the rear legs hard they dart through the water for short distances, which enables them to chase food or escape predators that remain behind on the seabed. When they walk, swimming crabs go sideways, like most other crabs. The reason for this lies in the construction of the leg joints. The whole body of a crab is covered in a hard, rigid outer skeleton, and the legs are so close together that there is no room for them to move forwards or backwards. They can therefore only move up and down. This up and down movement, together with a lateral stretch, leads to the sideways gait. Only in spider crabs, which have their legs more widely spaced, is some degree of diagonal movement possible.

SWIMMING CRAB

Right: The rear legs of the velvet swimming crab are adapted to form swimming 'paddles'.

Below: The red eyes may serve to indicate the velvet swimming crab's vicious nature.

VELVET SWIMMING CRAB
Liocarcinus puber

Size:	Up to 8cm across shell
Colour:	Brown body, blue joints and red eyes
Habitat:	Rocky seabed; low shore to 80m

Velvet swimming crabs are encountered quite often since they are sometimes found on the lower shore. The most ferocious of Britain's crabs, they have vivid red eyes and blue joints to the legs. They also have a fine coating of hairs that feel and look like velvet when clean, although they often accumulate fine silt and then simply look muddy. When challenged they try to make themselves look larger and more dangerous by stretching their claws out wide and they will often attack an aggressor regardless of its size. They have also been seen to attack one another at times, but whether this is over food or territory, or simply due to innate competitiveness, is not known.

During the day velvet swimming crabs tend to linger among boulders, in rockpools or in the holdfasts of the kelp forest. Here they feed using a number of methods. They may sift detritus, pick away at brown seaweeds or hunt other animals. To do this they take advantage of the 360° vision afforded by their stalked eyes. Alert and watching carefully, they remain motionless with claws outstretched until an unsuspecting animal, perhaps a small fish, comes within reach and is grabbed and eaten.

Facing page: Rocky submarine cliffs containing cracks and fissures serve as a suitable home for the velvet swimming crab.

Above: The colouring of the velvet swimming crab is distinctive, with its vivid red eyes and blue leg joints.

Right: Claws always appear to be at the ready.

VELVET SWIMMING CRAB

Left: Feather stars and brittle stars are often found together.

Inset: It is easy to see how feather stars got their name.

Right: In a dense bed of feather stars only the arms can be seen.

Feather stars prefer cool water and are only found in very sheltered places such as sea lochs to the north of the country. They often live at a considerable depth and are found in large aggregates attached to vertical or overhanging rock faces. They are the most primitive type of echinoderm found around Britain and there is only one species commonly found in our coastal waters. They retain a number of features reminiscent of their fossil ancestors; for example the mouth is on the upper surface rather than the lower as is the case in starfish, brittle stars and urchins. They also start life attached to rocks, and when they do eventually break free they still retain some 25 small, claw-like stalks,

FEATHER STAR	*Antedon bifida*
Size:	Up to 15cm in diameter
Colour:	Variable; usually brick-red
Habitat:	Rocky shore and muddy seabed; low shore to 200m

called cirri. It is with these cirri that they cling on while raising their arms to filter food from the sea. They move by crawling gracefully on their ten feathery arms and at times large numbers of them can be found high in the kelp forest. This position, far from the seabed, is an ideal location from which to catch passing plankton.

FEATHER STAR

Left: Luidia ciliaris *has seven rather than the usual five arms.*

Right: This young Luidia *is less shapely than larger specimens.*

Luidia *have a slightly patterned, softer-feeling body than most starfish.*

Luidia ciliaris	
Size:	Up to 40cm across
Colour:	Orange above, pale below
Habitat:	Sandy, muddy seabed; very low shore to 150m

Most starfish have five arms but *Luidia* always has seven. Each arm is slightly flattened and the tube feet, which are found in the groove on the lower side, differ from those of the common starfish in that they end in small knobs, not suckers.

Luidia are found in areas of coarse sand and gravel, where there is little water movement and silt has accumulated. Moving over the surface they hunt for their food, which mainly consists of other echinoderms.

LUIDIA

Astropecten irregularis are well adapted to living buried beneath sand or gravel. Their five arms are flattened and edged by two layers of marginal plates. The upper plates have one or two small spines, while the lower ones have a larger one. These spines, together with the pointed suckerless tube feet (among which the segmented worm *Acholoë astericola* often lives) are ideal for digging into soft sediments but mean the animals are unable to climb rocks in the way the common starfish can.

A. irregularis are carnivores and eat a wide range of sand-dwellers including worms, crustaceans and brittle stars as well as other starfish. The food is first grabbed by the tube feet and then pushed into the mouth whole, digested, and any remains later spat out.

These starfish maintain contact with the surface, and clear seawater, by projecting the tips of their arms up through the sediment. To respire, they create a current of clean, well-oxygenated seawater by vibrating the long spines which fringe each arm. Living beneath the sand it would be difficult for respiration to take place by other means: the gills would soon become clogged.

Astropecten irregularis	
Size:	Up to 12cm across
Colour:	Orange above, pale below
Habitat:	Sandy, gravely seabed; very low shore and shallow sea

Left: Astropecten irregularis *prey on other starfish.*

Below: A star-like depression shows where this starfish had buried itself.

ASTROPECTEN

Cushion stars are often inconspicuous, blending well with their surroundings.

Like a pin cushion in shape, the cushion star or starlet, as it is sometimes known, has rounded tips to its five short, broad arms. It is the smallest British starfish and is only found to the west and south-west of the country, with occasional sightings as far north as the Isle of Man.

An inconspicuous dull olive in colour, it adheres closely to the undersides of rocks, among which it searches out its food, a somewhat unpleasant-sounding diet of dead and decaying micro-organisms, plants and animals. The cushion star does, however, have a fascinating sex life. At about 1cm in diameter and two years of age, the animals are males. Later, when they have doubled in size to 2cm and have reached an age of four years, they then become functional females. Each female will lay about 1,000 eggs on the underside of a rock in May. These then hatch as young starfish after about three weeks.

CUSHION STAR	*Asterina gibbosa*
Size:	Up to 5cm in diameter
Colour:	Variable; olive-brown, green, yellow
Habitat:	Rocky seabed and rockpools; low shore to 100m

CUSHION STAR

BLOODY HENRY	*Henricia oculata*
Size:	Up to 10cm
Colour:	Blood red to purple above, pale below
Habitat:	Rock and gravel seabed; very low shore and shallow sea

Superficially Bloody Henry look similar to common starfish. They have five arms tapering to points from the central disc and orange to blood red coloration. They do however differ in a number of ways. The outer skeleton of Bloody Henry feels brittle and chalky, not leathery like the common starfish's. They are only rarely found on the seashore, preferring deeper rocky areas of the seabed. In addition, although they can feed by stomach eversion, like both the common and the spiny starfish, and are known to eat small hydroids, sponges and the like, they are also suspension feeders. To feed in this way they extend their arms up into the surrounding seawater. They then wave them about so that small food particles which are floating by are caught on strands of mucus and passed into the mouth.

Right: Bloody Henry starfish usually frequent deeper, rocky areas of the seabed.

Above: This Bloody Henry has a malformed arm, possibly due to regrowth after damage.

Below: Vivid scarlet is one of the two colours of Bloody Henry.

BLOODY HENRY

Top: Sun stars have a rough surface.

Bottom: Patterned red and white sun stars are common on rocky seabeds.

Unlike most starfish, the sun star does not have five arms but 8–13 blunt, rather short rays which extend from the central disc. The whole of the upper surface is covered with regular rows of stubby spines, which help to protect the animal from predators.

Sun stars are quite common on rocky seabeds in cool northern waters and are occasionally seen intertidally on rocks on the lower shore in the north. They cannot tolerate high temperatures and will quickly die if exposed to the air on a warm summer day.

Sun stars are predatory by nature. They frequently eat other types of starfish, especially the common starfish and will, if the need arises, follow their quarry from rocks on to sand or among mussel and oyster beds.

SUN STAR	*Crossaster papposus*
Size:	Up to 25cm across
Colour:	Brick-red to orange
Habitat:	Rocky shore and over shellfish beds; between 10 and 40m below low-tide level

SUN STAR

Sucker feet allow starfish to move easily over most surfaces.

COMMON STARFISH
Asterias rubens

Size:	Normally up to 15cm across, may reach 50cm in deep water
Colour:	Orange/brown
Habitat:	Rocks, boulders and over shellfish beds; low shore to 400m

The common starfish, with its five orange arms drawn out from the central disc, is familiar to most people. They are found on all British shores, especially among rocks. Starfish are able to withstand both reduced salinities (down to about 23 ‰) and considerable pressures, being found to a depth of about 400m.

Their upper surface is covered with small spines which make it feel a little rough. They also have small, soft finger-like extensions through which oxygen is taken up and waste products excreted. Finally, there are tiny jaw-like projections which keep the surface clean by catching and casting off any sediment or larvae that may settle onto it.

While starfish do not have 'faces' in the conventional sense, they do have eyes. These are quite simple and are found, not near the mouth, but at the end of each arm. Each eye is made up of a group of light-sensitive cells and the starfish regularly raises up the tips of the arms so that it can detect the direction from which light is coming. On the whole they prefer to escape daylight and live in dark, shaded places such as crevices and the undersides of rocks. However they, like many marine fish, appear to be attracted to a diver's flashlight at night. Why this should be remains a mystery.

Turn a starfish over and you will see how different the underside is. In the centre of the disc is the mouth and from it a groove extends down each arm. In each groove there are hundreds of small tube feet. Individually each foot has a small reservoir at the base and can act as an individual sucker. It is worked from

COMMON STARFISH

an interconnected fluid-filled hydraulic system that is operated through a valve on the top of the disc. Although these tube feet do not move together, they are controlled by the nervous system so that no matter which arm is temporarily leading the way, it gains dominance over the others and the animal only tries to go in one direction at a time! The tube feet are also used for a number of other tasks such as feeding and righting itself. The animal can loop its arms to such an extent that the tube feet grasp the substrate and pull the animal back over.

The common starfish is a voracious carnivore and is often found among mussel and oyster beds where it can constitute a serious pest. The method by which it feeds is quite extraordinary. Having decided on its prey, for example a mussel, it will surround it in a 'bear hug' with all the arms and pull on the two valves of the shell with the tube feet. Eventually the mussel is weakened and will gape slightly. At this point the starfish, which has no teeth, pushes its stomach out of its own body and into that of the mussel. It then releases digestive enzymes that kill the mussel and dissolve away its flesh. This rich liquid food, which contains little if any indigestible material, is then drawn back into the starfish together with the

stomach. Any bits of shell that may have been taken in are later spat out of the mouth.

Starfish have amazing powers of regeneration and put this ability to good use. If danger threatens they will often deliberately shed an arm, from the junction with the disc, a process called 'autotomy', rather than risk death. They can in fact lose all five arms and still regrow a complete set from the disc; alternatively, so long as there is at least one arm and half a disc, an entire animal may be regenerated. This process is, however, rather slow and may take up to a year to be successfully completed.

Like other echinoderms, common starfish are either male or female and fertilization of the eggs (as many as 2.5 million from each 15cm wide female) takes place in the open sea in spring and summer. To enhance the chances of success the starfish release pheromones (chemical messages) to stimulate others to release at the same time. When the young first hatch out in the sea they are rounded and look quite different from their parents. Only after growing for about three months in the plankton, followed by metamorphosis on a suitable substrate, do they take up the adult, bottom-dwelling, lifestyle.

Numerous common starfish can often be found in rockpools.

Top and bottom: The spiny starfish has prominent spines along its arms.

Spiny starfish are quite unmistakable because of their large size and prominent spines, which lie in rows along each arm. Each of the spines is surrounded by tiny pincer-like appendages called pedicellariae, which protect the animal and help to keep it clean by clearing away any detritus that lands on its surface.

Found to the south and west of Britain, spiny starfish are carnivorous and feed in a similar way to common starfish. They hunt for their food among the rocks and stones of the seabed and eat a wide range of animals, including crustaceans, molluscs (they are a considerable pest on shellfish beds) and other echinoderms.

In summer quite large numbers of spiny starfish are found together in aggregates. The sexes are separate and this communal breeding enhances the possibility of succesful external fertilization. Once an egg has been fertilized, it develops into a planktonic larva before maturing into the sedentary adult mode of life.

SPINY STARFISH	*Marthasterias glacialis*
Size:	Up to 25cm, occasionally reaching 60cm across
Colour:	Variable; grey, mauve, blue/green or yellow
Habitat:	Rocky, stony seabed; very low shore to 180m

SPINY STARFISH

Above: Dense 'carpets' of
brittle stars may occur
on sandy or gravel
seabeds.

Many colour variations
occur in common brittle
stars, this being one of
the more vivid.

COMMON BRITTLE STAR
Ophiothrix fragilis

Size:	Disc to 2cm diameter
Colour:	Variable; includes red, brown, purple and blue often banded
Habitat:	Rocky, sandy seabed; very low shore to 350m

B rittle stars are usually only found on the very low shore during spring tides, under rocks and boulders, as occasional individuals. However in areas of the seabed well below the low-tide level, they live either among other filter feeders in between, for example, the holdfasts in the kelp forest, or in areas where the sediment is composed of gravel or clean sand, where they can literally carpet the bottom. As many as 10,000 per square metre have been recorded.

They are, as their name suggests, very easily damaged, even when gently handled. It is also known that they can throw arms off by a type of self-amputation. Fortunately, like many starfish, regeneration of arms takes place both readily and rapidly. Both the arms and the central disc are beautifully coloured and it is

BRITTLE STAR

quite usual for alternate bands of colour to run down each of the five spine-covered arms. The arms are made of jointed plates, which give them great flexibility and allow them to cross the seabed speedily, each arm moving in a snake-like fashion.

Although some brittle stars are scavengers, eating dead animals, or deposit feeders, raking food into the mouth, the common brittle star is an omnivorous suspension feeder. To catch its food it raises one or more of its arms up above the seabed. This enables it to capture passing detritus or planktonic organisms on strings of mucus hung between the spines. This rich harvest is then moved by the tube feet to the mouth and eaten.

Unusually, these brittle stars are on an edible urchin, which normally removes anything.

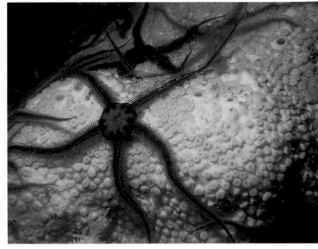

Above: In common with other starfish, brittle stars can regenerate missing legs.

Right: Black brittle stars are slightly bigger than the common brittle star.

Above: A sponge makes a contrasting backdrop for these brittle stars.

BLACK SEA URCHIN	
Paracentrotus lividus	
Size:	Up to 6cm in diameter
Colour:	Variable; black, purple, dark green and dark brown
Habitat:	Rocky seashore ; midshore to 30m

Black urchins only extend their range north to the Channel Islands, the west coast of Ireland and the Hebridean Isles. They are unique among urchins in that they have the ability to burrow into rock. They only do this in areas of the coast where the breaking waves could knock them off the smooth rocks. In such places they find a small crack which they then widen and deepen, using their teeth and spines, until they have formed a circular burrow in which to shelter.

Above: Black sea urchins are found only in a few areas on the west coast.

Right: Burrowing into rock is a characteristic unique to the black sea urchin.

BLACK SEA URCHIN

Green sea urchins can cover themselves with pieces of shell and pebble.

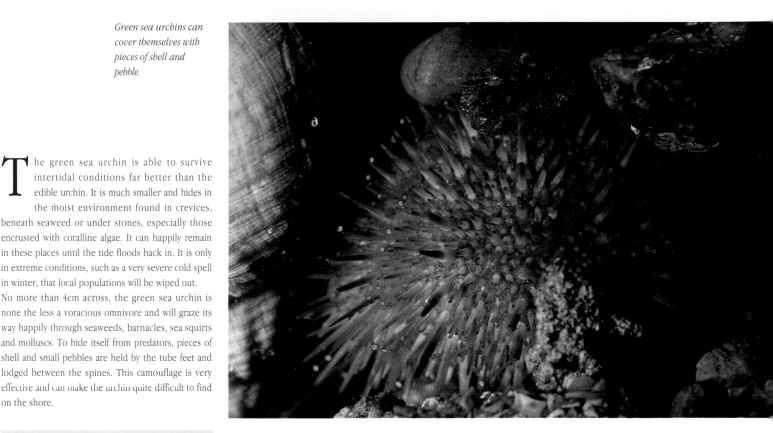

The green sea urchin is able to survive intertidal conditions far better than the edible urchin. It is much smaller and hides in the moist environment found in crevices, beneath seaweed or under stones, especially those encrusted with coralline algae. It can happily remain in these places until the tide floods back in. It is only in extreme conditions, such as a very severe cold spell in winter, that local populations will be wiped out.

No more than 4cm across, the green sea urchin is none the less a voracious omnivore and will graze its way happily through seaweeds, barnacles, sea squirts and molluscs. To hide itself from predators, pieces of shell and small pebbles are held by the tube feet and lodged between the spines. This camouflage is very effective and can make the urchin quite difficult to find on the shore.

GREEN SEA URCHIN
Psammechinus miliaris

Size:	Up to 3.5cm in diameter
Colour:	Slate grey/green with purple tips to the spines
Habitat:	Rocky shore; midshore down to 100m

GREEN SEA URCHIN

EDIBLE SEA URCHIN
Echinus esculentus

Size:	Up to 17cm in diameter
Colour:	Red, purple, pink or white, at times with purple tips to spines
Habitat:	Rocky shore and on kelp; below low-tide level to 50m. Has been recorded at 1200m.

This is the largest of the European sea urchins and like starfish it is pentamerous (divided into five parts). If you look closely you can actually see the five rays running from the top down to the underside. Each year as the animal grows there is a slight variation in the pigments deposited in the shell. From this it is possible to age these animals and they are known to live for up to twelve years.

The whole shell is covered with sharp, calcareous spines. This feature has given the entire phylum in which sea urchins are placed the name Echinodermata ('spiny skin'). The spines are important because, with a ball and socket joint at the base, they help the animals both to move and to protect themselves from predators. In between the spines there are tube feet, just like those of a starfish. These are used as suckers to move the animals about as well as playing a role in excretion and breathing. In addition to the spines and tube feet there are little stalk-like extensions called pedicellariae. Each of these has three 'jaws' at the tip and their job is to keep the urchin's surface clean by removing bits of dirt or larvae that have settled out from the seawater.

Edible sea urchins are nocturnal by nature and dislike any form of strong light. They are omnivorous and graze algae and encrusting animals off rocks or kelp fronds. The mouth is on the underside and is made of a group of five chisel-like teeth which are called Aristotle's lantern. When conditions are good and there is plenty of food available, large numbers of these urchins can be found together. Estimates of as many as 2,000 per hectare have been made for parts of the English Channel.

This urchin is called edible but in fact it is only usual for the reproductive organs, that is to say the caviar, to be eaten. Although rarely on offer in Britain today, they have been eagerly sought after around the Mediterranean for thousands of years, leading to a considerable reduction in the population.

In this sea urchin the sexes are separate and it is thought that a large female may release up to 20 million eggs between March and May. Once fertilized, the young grow for about two months among the food provided by the abundant summer plankton. They then settle down as tiny sea urchins, 1mm diameter, to a new life on the rocky seabed.

EDIBLE SEA URCHIN

Facing page: This close-up of the edible sea urchin shows the sucker feet and spines.

Right: The spines serve as an excellent deterrent to potential predators.

Below left: The sucker feet play a role in excretion and breathing, as well as being the urchin's method of locomotion.

Below right: The edible sea urchin's mouth has five 'teeth' which form the Aristotle's lantern.

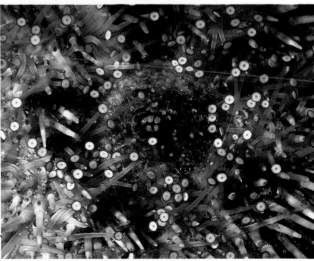

Facing page: The test of the sea potato can often be found along the tide line on sandy shores.

This page. Tracks in the sand show the meandering route taken by a sea potato as it feeds.

The sea potato, or heart urchin as it is sometimes called, is somewhat unusual in that it lives not on rocks but buried in sand. It is quite common on all sandy beaches, but most often it is the remains of the delicate shell on the strandline that is first seen. Sea potatoes burrow at a depth of about 20cm on the low shore. The sand protects them from predators and because of this, their shells are far more fragile than those of the surface-dwelling urchins. Other adaptations to life beneath the beach include short, flattened, paddle-like spines that can dig sand, and the ability to maintain contact with the surface by the extension of a number of very long tube feet through a narrow vertical funnel.

Because of their shape, most urchins and starfish can move in any direction. Sea potatoes are more oval in shape and have a definite 'head' and 'tail'. Their golden-yellow spines all lie at an angle away from the front so that the animals can move easily between the grains of sand. They are deposit feeders and must constantly burrow if they are to survive by collecting particles of food with specially adapted, prehensile tube feet, in addition to their mouths.

SEA POTATO	*Echinocardium cordatum*
Size:	Up to 9cm long
Colour:	Cream
Habitat:	Sandy seabed; lower shore to 200m

SEA POTATO

The lesser spotted dogfish can sometimes be seen among the seaweeds of a rocky shore where the female lays her eggs.

LESSER SPOTTED DOGFISH	
Scyliorhinus caniculus	
Size:	Up to 75cm
Colour:	Dark brown spots on sandy brown background
Habitat:	Sandy, gravelly or muddy seabed; shallow sea to 50m, occasionally to 110m

The greater spotted dogfish forages for molluscs and crustaceans along the sandy seabed.

GREATER SPOTTED DOGFISH
(NURSE HOUND OR BULL HUSS)
Scyliorhinus stellaris

Size:	Up to 150cm
Colour:	Dark brown spots on sandy or grey background
Habitat:	Rocky and sandy seabed to south and west of Britain; shallow sea to 60m

Dapple-brown dogfish are the best known of British sharks. They are found in shallow seas all round the country, although the greater spotted dogfish is most common to the south-west.

Both types are sandy brown in colour with dark brown spots. This coloration camouflages them well against the sandy seabed over which they hunt. They mainly forage over the seabed taking molluscs and crustaceans such as crabs, lobsters, shrimps and prawns and occasionally flatfish, gurnard and dragonet. At times they will also feed in midwater where they catch fish such as herring, pilchard, pouting and whiting. They have a big stomach which enables them to take large, if irregular, meals. Once they have fed, they tend to stop swimming and lie still, either on their own or in groups, their protective coloration helping them to merge into the background and avoid predators.

Like all sharks dogfish have no bones, only a tough cartilaginous skeleton. The only really hard parts are their teeth which are all the same shape, inward pointing and sharp enough to cut and tear food apart. The skin of the fish also feels rough because it too is covered by small 'teeth' called denticles. These, if stroked the wrong way, are sharp enough to graze or cut the skin. This property has been used for centuries because man soon learnt that, when dried, dogfish skin was a marvellous natural sandpaper or non-slip sword grip.

DOGFISH

The only time you are likely to encounter a dogfish on the seashore, or caught by the tide in a rockpool, is between November and July, when the females are laying eggs. Mating has usually taken place in deeper water in late summer. When she is ready to lay her eggs, usually about twenty each year, she swims from her deeper-water, sandy habitat, to the seaweeds on the rocky shore. Here she swims round the weed so that as each egg (mermaid's purse) is released it becomes entwined by the long rubbery tendrils at each corner. It is important that they are lodged firmly because it takes 8–10 months, depending on the water temperature, for the eggs to hatch. The young of the greater spotted dogfish are about 16cm long, those of the lesser spotted dogfish only about 10cm. Both may hatch with the yolk sac still attached but as soon as they start to feed this drops off.

Of the two, it is the lesser spotted dogfish that is more commercially valuable. It is caught in large numbers and is sold, often to the European market, as rock eel or rock salmon.

Like many marine dwellers, the dogfish has a strangely shaped eye.

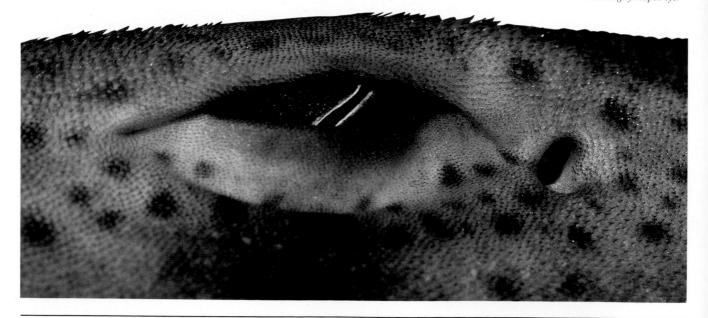

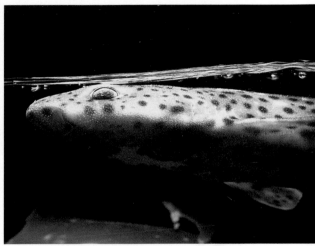

Commonly known as mermaid's purse, the dogfish egg case (left) takes 8-10 months to hatch into a young dogfish like the one pictured above.

The smooth hound is easily recognized as a member of the shark family.

SMOOTH HOUND	
Mustelus mustelus	
Size:	Up to 120cm
Colour:	Slate grey above, cream below
Habitat:	Sandy seabed; shallow coastal sea to 100m

STELLATE SMOOTH HOUND	
Mustelus asterias	
Size:	Up to 150cm
Colour:	Slate grey above with numerous white spots over the back and sides, pale beneath.
Habitat:	Sandy and muddy seabed; shallow coastal sea to 150m

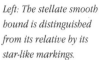

Left: The stellate smooth hound is distinguished from its relative by its star-like markings.

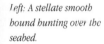

Left: A stellate smooth hound hunting over the seabed.

There are just two species of smooth hound found around Britain, mainly to the south and west of the country. Although they look like many of the world's more ferocious sharks, they are in fact rather timid. They spend the winter in relatively deep water and then move inshore in spring to hunt over shallow sandy seabeds for molluscs, bottom-living fish, crabs and other crustaceans. These they eat by grinding them between their small, flat, button-like teeth.

The main difference between the two types of smooth hound, apart from the starry markings, is the way in which their young develop. In the smooth hound the female will bear about fifteen young, each nourished within her by a placenta. Eventually when they have matured they are born at about 30cm in length. In contrast, the stellate smooth hound may have as many as 30 young at a time but although they are the same size when they hatch they develop with a yolk sac, rather than attached to the mother by a placenta.

SMOOTH HOUND

THORNBACK RAY *Raja clavata*

Size: Body of male up to 85cm,
              female up to 120cm
Colour: Sandy brown above, pale
              below
Habitat: Sandy seabed; shallow sea
              to 100m

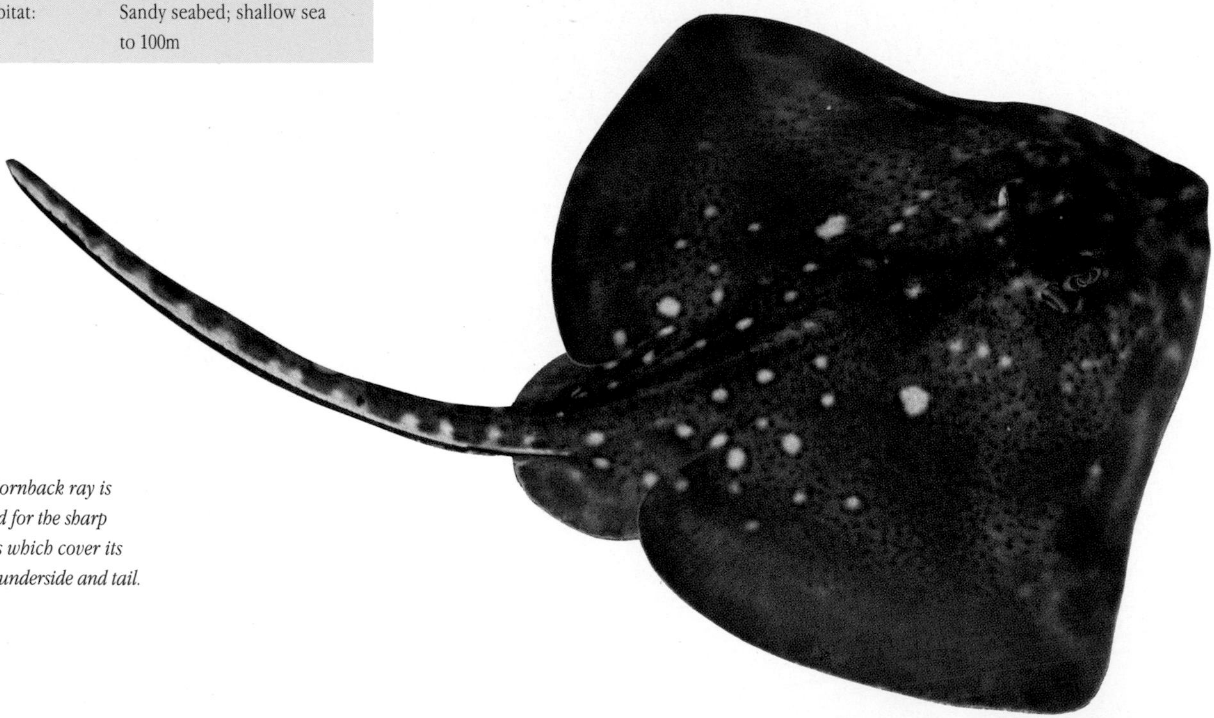

*The thornback ray is
named for the sharp
thorns which cover its
back, underside and tail.*

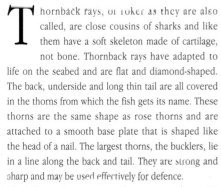

Eyes are above and the mouth below the body of the ray.

Thornback rays, or roker as they are also called, are close cousins of sharks and like them have a soft skeleton made of cartilage, not bone. Thornback rays have adapted to life on the seabed and are flat and diamond-shaped. The back, underside and long thin tail are all covered in the thorns from which the fish gets its name. These thorns are the same shape as rose thorns and are attached to a smooth base plate that is shaped like the head of a nail. The largest thorns, the bucklers, lie in a line along the back and tail. They are strong and sharp and may be used effectively for defence.

The skin is pigmented on the upper surface with numerous small black spots over a grey/brown background, giving it a patchy appearance when seen out of the water. In its natural habitat on a pebbly, muddy or sandy seabed it is camouflaged extremely well. As with all flatfish the lower surface is not coloured because it is usually on the seabed and hence rarely seen. Rays swim by gently waving the edge of the wide pectoral fins in the form of an 'S'. The wave starts at the side of the head and passes along the margin of the fin towards the tail. The tail itself has little power to propel the fish and simply acts to direct and stabilize it.

Most fish breathe through their mouths, with the water passing over the gills and out through the gill arches or operculum. This would cause problems for the ray because it would frequently take in a mouthful of sediment that would then clog the gills. However, the problem has been overcome. Two small holes called spiracles, which in the sharks are seen behind the last gill arch, have moved round, one behind each of the ray's eyes. They can be seen opening and closing, with the clean water passing in over the gills and out through the openings on the lower surface. Thornback rays have good eyesight. The large

THORNBACK RAY

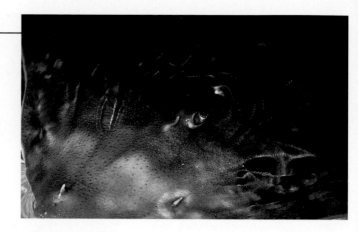

movable eyes are raised up so that they get a good all-round view. However, because they are positioned well back on the upper surface the animals cannot in fact see the food they are eating. Instead they use their highly developed senses of touch and smell. The nostrils are found on the lower tip of the snout and are covered with flaps. Rays also have an additional sense: the ability to detect differences in the Earth's magnetic field around them. The detection takes place in tiny pores found around the mouth and the information allows them to find their way about, and hunt their prey, in dark and turbid water.

Their food mainly consists of bottom-living animals which they take from the seabed. Thornback rays usually pounce, bite and then grind the food up using their flattened teeth. In the young, the diet tends to be restricted to small fish and shrimps. In more mature individuals it is more varied and includes flatfish, sandeels, crabs, ragworms, hermit crabs and other shellfish. As adults they also make excursions

Above and below: Although the thornback ray has relatively good eyesight, it relies on highly developed senses of touch and smell to identify its prey.

away from the seabed and may eat herrings, sprats and gadoids in considerable numbers.

As is typical of the family, male thornbacks have two rod-like claspers and fertilization is internal. The females migrate inshore in winter and lay about 20 eggs each. Every egg is laid in an individual leathery rectangular dark case called a mermaid's purse. Each

is about 7.5 X 6cm in size and tends to be flat on one side and more rounded on the other. Each of the four corners extends to form a small hollow tendril and the margins are covered with fine soft hairs which help to attach the egg to seaweed or rocks. It takes about 4–5 months for the eggs to develop and as a result the young rays emerge when food is most plentiful.

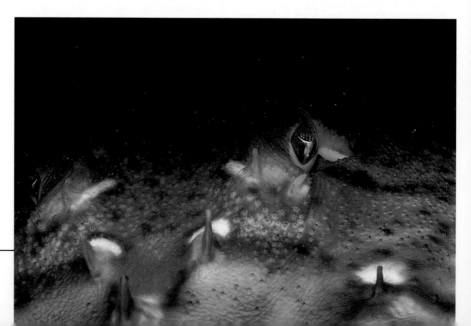

Ferocious-looking conger eels probably don't deserve their reputation for being vicious.

Conger eels are quite unmistakable because of their snake-like profile and sheer size. There are records of fish as much as 270cm in length and 73kg in weight. Their long body is smooth and scaleless and single dorsal and anal fins run along the upper and lower margins, joining at the tip to form a continuous fringe around the pointed tail.

Conger eels are found on all Britain's rocky coasts. They usually live in fairly shallow water and may even be found between the tides or in large rockpools. Unlike the common eel they only live in the sea and never enter freshwater. Because of the flexibility of the body, which results from the large number of vertebrae (there may be over 100), conger eels can get into quite suprisingly small nooks and crannies along a rocky or boulder-strewn coastline. They are nocturnal and will remain in their chosen crevice until nightfall.

Large-eyed and with a very good sense of smell, congers can be voracious night predators. They are

CONGER EEL

unfortunate in that they have gained a probably unwarranted reputation for viciousness. Their normal diet consists of bottom-living fish like flatfish, dogfish and rockling, shellfish like crabs, lobsters and octopus, and fish from more open water such as herring, cod and pollack. Conger eels have rows of small, backward-pointing, needle-like teeth in both the upper and slightly protruding lower jaw. In fact, rather than biting or chewing their food they tend to

swallow it whole, sucking it in in one, rather noisy, action. Once they have fed well they return to their crevice and may spend a day or two digesting the food before setting out to hunt again. With such a rich and varied diet conger eels grow rapidly and may weigh as much as 40kg by their fifth year.

The life cycle of the conger eel is particularly interesting. They become sexually mature at between five and ten years of age, spawning only once. As they

come into breeding condition a number of radical changes take place. They stop feeding and lose their teeth, the gut and other organs start to degenerate and the skeleton becomes decalcified. At the same time the gonads increase in size to fill the body cavity, accounting for about a third of the body weight. The fish has effectively become a mobile egg or sperm case. This dramatic transformation takes place in the summer and the eels swim south to an area between the Azores and Gibraltar. Here, in midwater at a possible depth of between 2,000 – 4,000m the eels spawn and die. Each female may have laid between three and eight million eggs. These drift freely and once hatched the larvae (known as leptocephalus) move passively north-eastward on the North Atlantic Drift current. After a year or two, when quite close inshore, they metamorphose into what we recognize as conger eels.

This page: The conger has rows of small, needle-like teeth in both the upper and lower jaw.

Facing page: With its large eyes and good sense of smell the conger is an efficient night predator.

CONGER EEL	*Conger conger*
Size:	Up to 2.7m
Colour:	Uniform grey above, paler below
Habitat:	Rocky seabed and wrecks; shallow sea to 100m, to 3,000–4,000m when spawning

*Facing page and right:
The bib, or pouting,
prefers warmer, inshore
waters with shallow,
sandy seabeds.*

BIB (POUTING)	*Trisopterus luscus*
Size:	Up to 30cm, long chin barbel, larger lower jaw
Colour:	Copper with vertical banding
Habitat:	Rocky and sandy seabed; coastal and open seas to 100m

A s a group of fish the cod family is best known because of its commercial value. All are edible and many have been the main catch of fishermen throughout Europe and eastern America for centuries. Cod was the most important and was used traditionally fresh, dried, salted or smoked so that it could feed the community all year round, even when the weather was too bad for the fishing fleet to go to sea. Today the cod-related fishing industry remains the most important in the world, accounting for about ten million tons per annum (17 per cent of the world's fish catch) an

amount only exceeded by landings of herring which are far less valuable.

The more specialized fish in the cod group are all characterized by the presence of three dorsal and two anal fins, which are made of soft rays. On the whole their skin is covered in smooth, small scales.

Fish in this family are found throughout European and North Atlantic waters. Cod range as far as Newfoundland and the eastern seaboard of the USA. Saithe are most common in inshore northern areas and are often found around Hebridean piers and rocky coasts. Bib prefer warmer, more southern,

inshore, shallow, sandy coasts, and whiting are found in midwaters from the Bay of Biscay to Iceland. They are all shoaling fish and usually live close to the seabed. Although there are no real physical boundaries in the sea, many of the fish live in fairly discrete populations. Each group of cod for example, has its own particular yearly migratory pattern to feeding and spawning grounds. As a result, once the main areas in which they lived were discovered in the nineteenth century, they could be, and were, heavily fished.

At first some of the cod caught were huge, weighing

THE COD FAMILY

as much as 90kg, and must have been some considerable age. In contrast to this a cod of 18–20kg is considered large today and most of those caught commercially only average about 4.5kg. Many of the other fish in the group are naturally much smaller. Whiting, the most common, rarely exceed 40cm and bib only occasionally reach 30cm.

Living close to the seabed, much of the diet of these fish includes shrimps, crabs and other crustaceans, worms and bottom-living fish such as sandeels and gobies as well as fish of the open sea such as herring and caplin. Because of the nature of their habitat and habits they do not need to swim continuously and at times remain suspended, motionless, as a result of the bouyancy afforded by their swim bladder and oil-rich liver. Much of the time fish such as cod, which prefer cool water of less than 10°C, have to feed in the dark or in turbid water, and in such places their eyesight is of little use. In this situation they use their highy sensitive chin barbels to explore the seabed for food. Once something suitable has been found the strong, sharp teeth, set within powerful jaws, can seize and crush the prey.

All the members of the cod family breed between late winter and spring. The age at which they become sexually mature varies from one year in the bib to

COD	*Gadus morhua*
Size:	Up to 1m, long chin barbel, larger upper lip
Colour:	Dapple brown, pale curved lateral line
Habitat:	Sandy, muddy seabed; shallow sea to 600m

POLLACK	*Pollachius pollachius*
Size:	Up to 1.2m, no chin barbel, larger lower jaw
Colour:	Uniform grey, dark curved lateral line
Habitat:	Rocky seabed; shallow and open sea to 100m

Cod share the barbel on their chin with other members of the cod family.

three years in the cod. Cod are extremely prolific and a large female may lay as many as nine million eggs. This in itself should go some way to helping the fish stocks recover, so long as fishing quotas and catch sizes are carefully regulated. In all the species the eggs and milt are simply shed together into the sea. The mature fish within the shoal all release at the same time to maximize the chances of successful fertilization. In the cod this is helped by the fact that the fish grunt to one another to indicate their readiness to spawn. These grunting noises have been recorded and played back to the fish, which are immediately attracted to the source of the sound. It would appear that in the wild the grunting noises bring fish together and synchronize spawning.

The fish lay small eggs which float freely in the the sea and are widely distributed away from the adult spawning areas by the currents. After two or three weeks they hatch and the larvae start to eat plankton and grow. Young cod remain in the surface waters for about two months and then, when they are only about 2.5cm long, they descend to the seabed where they feed on small crustaceans. Whiting are fish of more open midwaters and although they spawn at a depth of about 100m the young ones, which are up to 3cm long, are often found together in small shoals around the coast. It is not uncommon for them to swim along in front of, and if danger threatens seek shelter beneath, the tentacles of jellyfish such as *Cyanea lamarcki* and *Chrysaora isosceles*. Since the tentacles of these jellyfish have stinging cells, it is thought that the young whiting must have developed at least a degree of immunity to the poison. The young of all the different species are relatively easy to find around the coastline in summer. They are usually found in rocky areas and around piers and in harbours. In such shallow, warm, calm locations the shoals of young fish find ample shelter and grow rapidly on the rich variety of food.

The pollack lacks the cod's barbel.

Whiting are found in more open midwaters and spawn at a depth of about 100m.

SAITHE (COLEY)	*Pollachius virens*
Size:	Up to 70cm, minute chin barbel, larger lower jaw
Colour:	Olive grey with straight, pale lateral line
Habitat:	Rocky seabed; shallow sea to 200m

WHITING	*Merlangius merlangus*
Size:	Up to 70cm, minute or absent chin barbel, larger upper jaw
Colour:	Silvery blue/green, paler below, dark spot by pectoral fin
Habitat:	Rocky seabed; shallow sea to 200m

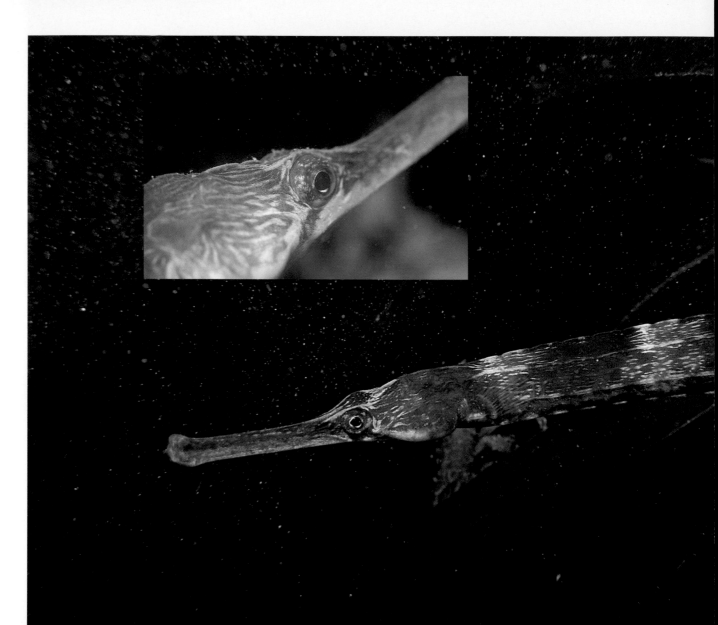

WORM PIPEFISH	*Nerophis lumbriciformis*
Size:	Females up to 17cm, males up to 15cm, no pectoral or anal fins, 63–72 body rings
Colour:	Variable; plain or dappled green and brown
Habitat:	Sheltered rocky shore among weed; low-tide level to shallow sea

which are wrapped over the eggs. It is about five weeks before the little ones (about 3cm in length) hatch and for a while they may rush back to the safety of the pouch if they sense any form of danger. With worm and snake pipefish, however, the eggs are simply glued to the shallow depression along the belly. The young, having hatched between June and August at 1cm, then spend a month or two in the plankton where they grow to around 3cm. By September they have settled out on to the seaweed-covered seabed and are quite common and easy to find in suitable areas.

The worm pipefish is smaller than its relatives.

One of the most distinctive fish in British waters is the John Dory. It is only really common west of the Isle of Wight, in the Irish Sea and to the south. Fish caught further north or east have usually been carried there by the Gulf Stream and they are usually somewhat sluggish as a result of the colder water.

The shape of the John Dory is unusual. The body is flattened from side to side and the dorsal fin has strong, tall spines, each covered in a long filament of tissue. When frightened or excited the fish raises the spines to make itself look larger and more threatening. The skin is covered in dapple markings, the intensity of which changes depending on the mood of the fish and the environment in which it is found.

In Italy these fish are called *janitore* – the janitor – and in France *jaune dore*, referring to their golden yellow colour, and it is from these that the name John Dory is thought to have come. On each side of the body is a large dark eye spot and these are reputed to be the thumb and fingerprint of St Peter, left after he took the tribute money out of the fish's mouth – *Pull up the first fish you hook, and in its mouth you will find a coin worth enough for my temple-tax and yours.* (Matthew 17:27). It is from this account that the Welsh name *pysgod pedr* has come.

John Dory are poor swimmers and tend to maintain their position by weak undulating movements of the dorsal and anal fins. Usually they are solitary fish but at times they form small shoals of four or five individuals, living near the seabed among kelp, close to rocks or sheltering near the surface beneath driftwood.

When they are about four years old, John Dory become sexually mature. It is known that they can sometimes breed around our southern coasts. However, they have to wait until July or August, when the sea temperatures are reaching 16–17°C, to be successful and even then the larvae develop far more slowly than they do in the Mediterranean, where they hatch between March and May. John Dory that have reached the North Sea are unable to breed because it is too cold, and they will eventually die as the sea temperature falls below 10°C with the onset of winter. These fish have excellent eyesight and frequently hunt at night. The binocular vision, which enables them to see things close to their faces, is directed forwards and is helped by the mobility of the eyes within the sockets. The John Dory stalks its prey with remarkable efficiency. Despite the fact that it is a poor swimmer it feeds on pilchards, sprats, sand eels, sand smelts, gadoids and many others, all of which are able to swim more quickly than it can. It catches its prey in an extraordinary way, making use of the fact that, because it is a very narrow fish, it is almost invisible to the prey it is approaching. It creeps up stealthily until it is close enough to strike. Suddenly the large jaws hinge forwards into an enormous protrusible tube. This action causes a large volume of water to rush into the mouth and with it the prey, which may be a fish more than half the length of the John Dory itself.

JOHN DORY	*Zeus faber*
Size:	Males up to 45cm, females up to 65cm
Colour:	Dapple brown body with large 'eye' on each side
Habitat:	Rocky seabed and open sea; shallow sea to 200m

Facing page: The John Dory is unmistakable.

JOHN DORY

Bass have long bodies covered in large, firmly attached scales. Along the back there are two dorsal fins, the first of which is made up of 8–9 sharp spines. Mature animals are metallic blue–black along the back with silver sides, while younger ones (up to 10cm in length) often have dark grey spots. There are also dark markings over the gill coverings and this area should be handled with care because it is covered with sharp, forward-pointing spines.

Bass are migratory fish. In October and November they move offshore to overwinter, returning to the coast in the spring. At this time they frequently enter estuaries, brackish water and even freshwater, and they are particularly attracted to warm-water outfalls.

They are voracious predators and put their numerous sharply pointed teeth to good use. During the summer months they usually feed along the shoreline, darting into the waves to devour any animals that have been churned up by the wave action. During the day these may include sandeels, sprats, worms, crustaceans and herring, whereas by night they will take bottom-living animals such as flatfish, worms and newly moulted crabs which they probably locate by their sense of smell. Younger fish are unable to catch such a wide variety of prey and are restricted to eating small crustaceans and fish.

After the age of five, bass move into estuaries and brackish water between May and August to breed. When first laid the eggs float freely in the surface water of the sea and hatch rapidly after a period of only four or five days. The young fish then tend to congregate among tidal pools and the creeks of salt marshes, which act as nursery grounds for their first year. Bass are very slow-growing and it can take five years for them to reach 0.5kg and 20 to reach 5kg.

They are a popular catch and because of this it is of vital importance not only that the spawning and nursery grounds are well protected and maintained but also that minimum legal catch sizes are closely adhered to. This should ensure that fish have reached reproductive age and spawned at least once before they are caught. They are a valuable catch with tender, firm, juicy, white flesh. They are also a great sport fish, usually taken with rod and line from the beach. The British rod-caught record stands at present at 8.334kg and this was probably a female, because they grow both more quicky and ultimately bigger than the males.

BASS	*Dicentrarchus labrax*
Size:	Up to 80cm
Colour:	Silver
Habitat:	Sandy, rocky seashore and estuaries; shallow sea to 100m

Sea bass have a really silvery appearance.

BASS

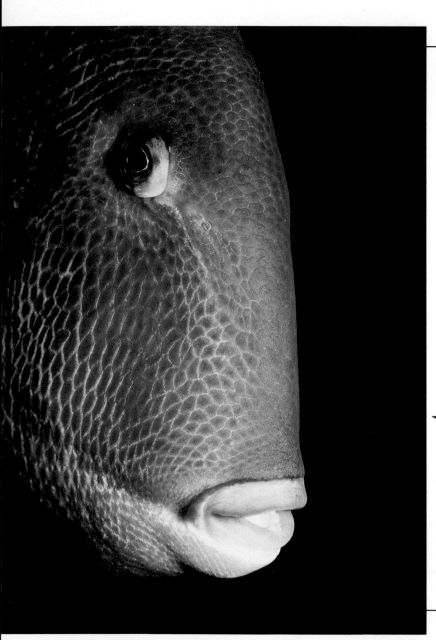

*The trigger fish is a
summer visitor to our
waters, carried by the
Gulf Stream from
further south.*

TRIGGER FISH	*Balistes carolinensis*
Size:	Up to 40cm
Colour:	Slate grey, brown
Habitat:	Rocky seabed and open sea; shallow sea to the south-west of Britain

Trigger fish are deep-bodied, oval and narrow. Their leathery, slate grey skin is covered with large diamond-shaped scales that frequently have a hint of iridescence when caught in the sunlight. They are not very common and are only occasionally caught to the south-west of Britain and Ireland or in the Irish Sea. This is because they are essentially warm-water fish, frequenting the Mediterranean and southern Europe. They swim rather weakly, and the young can be carried to our coast by the warm Gulf Stream on its northward path. They arrive fit and well but are destined for a one-way trip because as the sea temperatures fall with the onset of winter, it rapidly becomes too cold for them and they die.

Trigger fish get their name from their unusual dorsal fin which has been modified into three strong spines. The first is the largest. It is rough-edged and can be locked into position by a projection on the second smaller supporting spine. It is impossible to lower these manually. The third spine acts as a trigger to the others. It is the release mechanism and must be depressed before the first spine can be lowered. When lowered the spines lie neatly along a groove on the fish's back, in front of the second, normal dorsal fin. The spines and their trigger mechanism are used by

Rat-like teeth indicate the trigger fish's ability to crunch up crabs.

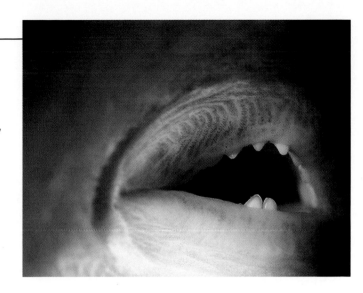

the fish for defence. At night trigger fish sleep, usually in a crevice in the rocks of the sea floor and such places are inherently dangerous for sleeping animals. To protect itself from hungry predators while it sleeps it wedges itself tightly into the crevice by raising its spines. If it is attacked it is almost impossible for a predator to dislodge it and the fish soon wakes up and retaliates by using the rough spines, which can inflict considerable damage.

Trigger fish are themselves active predators. Although they have small mouths, their teeth are large, rat-like and very strong, as is their jaw. This enables them to break open the shells of bivalves and crustaceans.

Their favourite food seems to be lobsters and crabs, which accounts for the fact that they are so often caught in lobster pots. Experience has taught them to be careful about attacking their prey. For example, when they have chosen a crab, they observe it closely, watching the claws move in and out in a characteristic threat display. Once the rhythm is determined, the trigger fish times its dive for the moment when the claws are apart, cleverly avoiding a nip. The rat-like teeth give an effective bite between the eyes, where the brain is located. The crab dies immediately and is then flipped over and the tail is opened and the meat eaten.

TRIGGER FISH

Left: Ballan wrasse
guard their eggs,
chasing off any fish
which approach too
close.

BALLAN WRASSE	*Labrus bergylta*
Size:	Up to 50cm
Colour:	Variable; mottled green, brown, rust, scales have yellow, green or brown bands
Habitat:	Rocky seabed; shallow coastal sea to 20m

Facing page: Cuckoo
wrasse seem almost
tropical in colouring.

Right: This ballan
wrasse is hovering in the
water waiting for the
shellfish to emerge.

CUCKOO WRASSE	*Labrus mixtus*
Size:	Males to 35cm, females to 30cm
Colour:	Females orange with three dark spots along dorsal fin, males orange with blue head, sides and tail
Habitat:	Rocky seabed; shallow coastal sea to 10m

Wrasse are found all over the world in both tropical and temperate waters. Seven species are found around Britain, four of which are common. They thrive on the rocky coast with its nutrient-rich water and good illumination, which provide the diversity of sessile plants and animals that the wrasse eat.

All four wrasse have a similar silhouette. They are laterally compressed, are covered in large scales and have one dorsal fin which runs almost the entire length of the elongated body. Two of them, the goldsinny and corkwing, are small, growing at most to 25cm. The cuckoo and ballan, on the other hand, may grow to 35cm and 50cm respectively. However, their most memorable feature is their coloration. They may never display the startling colours of their cousins, the tropical reef cleaner fish but they are, none the less, most eye catching.

They are all bright shades of brown, green, blue or rust, often with dappled, mottled markings. Of the four, however, it is undoubtedly the cuckoo wrasse which is the most colourful. Both the young and the breeding females are orange to red in colour and each has three dark spots towards the end of the dorsal fin. In contrast, the male has a vivid blue head which is overlaid with a mosaic of dark purple lines. The rest of his body is bright yellow or orange and his tail is edged with a wide band of a royal blue similar to that of the head.

All the wrasse have thick protruding lips and strong teeth, both in the jaws and on the pharyngeal bones. With these they are able to enjoy a mixed menu of shelled animals including crustaceans and molluscs. Such tough food may be ideal for the adults but is unsuitable for the larvae, which need to start life on a diet of plankton. To ensure that the eggs hatch in early summer, at a time when there is plenty of

WRASSE

Left and above: While the male cuckoo wrasse is one of the most colourful fish, the female is relatively drab.

CORKWING WRASSE
Crenilabrus melops

Size:	Up to 25cm
Colour:	Variable; olive, reddish or pale brown, spot by tail below lateral line. Breeding males have bluish lines on head and tail.
Habitat:	Rocky seashore amongst weed; coastal sea to 50m depth

Right: the corkwing is another colourful member of the wrasse family.

Goldsinny wrasse have a distinct black spot near the tail.

planktonic food in the sea, courtship starts in spring. Goldsinny wrasse spawn in moderately deep water, whereas in the other three species the male prepares a nest. This is built below the low-tide level from a collection of small stones. First he cleans them with his teeth, then he binds them together with seaweeds and mucus. Ballan wrasse simply clean a large expanse of rock face, on to which the eggs will be deposited and glued.

Once the chosen site has been fully prepared, the male will entice a female to enter the nest area and lay her eggs for him to fertilize. He will then guard and fan the eggs while they develop until they hatch and are dispersed within the upper reaches of the sea. In both ballan and cuckoo wrasse this complex breeding behaviour starts when the fish are about six years old. In any one stretch of coastline there will be a dominant male and if he dies the most senior female undergoes a sex change and becomes the next 'top' male. When breeding, the already colourful male cuckoo wrasse becomes even brighter so as to attract as many females as possible.

Since wrasse are slow-growing and long-lived (up to 20 years) a male may be able to hold his territory for some considerable time. Their longevity is also helped by the fact that, although they are an essential ingredient in French bouillabaisse, wrasse are not considered to be worth eating in Britain. As a result they are rarely subjected to the angler's lure. Indeed, as many an observant diver will testify, when left to their own devices they frequently settle down, leaning a little to one side, to sleep peacefully among the rocks, at times so soundly that they can be gently handled without disturbance!

GOLDSINNY WRASSE	
Centrolabrus rupestris	
Size:	Up to 18cm
Colour:	Uniform reddish brown, black spot near tail above lateral line
Habitat:	Rocky seashore; shallow sea to 30m

LESSER WEEVER FISH	
Trachinus vipera	
Size:	Up to 12cm
Colour:	Dapple grey/brown above, pale below
Habitat:	Sandy seabed; shallow coastal sea

Weever fish are rare examples of poisonous fish in British inshore waters. The two species found around our coast look rather similar, except for the fact that the greater weever fish may be twice the size of the lesser and it has two small spines on the front, upper edge of its eyes. Both have large oblique mouths and eyes that are raised so high that they are almost on top of their heads.

It is the position of the eyes that gives a clue to the fish's habitat. They are true sand dwellers and during the day lie quietly hidden waiting for their prey. They hide by rapidly burrowing in between clean sand grains by wriggling their bodies and their pectoral fins. Only the eyes and erect dorsal spines are left exposed above the sand. From this vantage point they await their food. In the lesser weever fish this is mainly shrimps, although they also take a wide variety of other marine creatures including small fish, shellfish and crabs. The greater weever fish also eats dragonets, gobies and small flatfish.

Weever fish spawn in the summer between June and August, the lesser weever fish usually moving a little offshore at this time. They shed small eggs (about 1mm in diameter) which float freely amongst the plankton while they develop.

Weever fish have venomous spines, used for defence and found on the first, black dorsal fin and the gill coverings. The spines have deep grooves which hold the poison after it has been produced by glandular tissue at the base. Since the greater weever fish lives in deeper water, it is the one more likely to be encountered by fishermen, whereas being common on or near sandy beaches the lesser weever fish is regularly found by the rest of us. If handled roughly or accidentally trodden on while paddling in shallow water, it will give an extremely painful sting, similar to that of viper. The effect may last for up to 24 hours and the pain may be felt not only at the site but throughout the entire limb. The venom attacks the blood corpuscles and the best action to take if stung is to let the wound bleed freely before cleaning it and then to seek immediate medical help.

Above: The weever is one of very few poisonous fish in our temperate waters.

WEEVER FISH

The common dragonet is most usually found on parts of the seabed made of soft material such as sand, shell or mud. Here they forage among the sediment for the variety of worms, crustaceans and molluscs that make up their diet.

Dragonets have rather long broad heads, with large raised eyes, small gill openings and low, wide, soft mouths. Together these give a somewhat frog-like appearance. However, it is their colouring that is their most distinctive feature. They all have smooth, rather slimy, scaleless skin and both females and immature males are well camouflaged, being dull brown and inconspicuous. It is the mature males that catch the eye. Their background colour is yellow, over which there are brilliant blue lengthwise stripes. In addition, both the dorsal and caudal (tail) fins have become extremely elongated in these males, the dorsal fin extending out like a long pointed sail.

Male dragonets grow larger than females but live for only five rather than seven years and it is thought that they may possibly only breed once at the end of their life. Spawning can occur at any time between January and August, but is at its peak between February and March. It is then that the male puts his dramatic shape and colour to good use. He undertakes an elaborate courtship dance, in which he puffs up his face and gill coverings to twice their normal size and swims around the female, while also driving off any other males. As he swims he raises his fins, flashing before her the striking iridescent blue bands. Once she has been attracted by his attentions they swim together in a nuptial dance, moving up towards the sea surface. They end by swimming so close to one another that their anal fins actually lock together to form a funnel into which both eggs and milt are deposited and where they mix before their release into the open sea. The eggs remain in the plankton where they mature and hatch and the larvae feed, grow and disperse. Eventually the young fish settle on to the sea floor when they are about 1cm long.

Dragonets are very common on sand.

COMMON DRAGONET	
Callionymus lyra	
Size:	Males to 30cm, females to 20cm
Colour:	Females and immature males dull brown, mature males brilliant yellow/blue
Habitat:	Sandy and muddy seabeds; shallow sea to 100m, occasionally to 400m

COMMON DRAGONET

Left: although able to swim properly, butterfish often hug the bottom.

Butterfish, or gunnel as they are also called, are members of the blenny family and although they have a small head they have the same characteristic thick fleshy mouths and large eyes. Their bodies are long and eel-like as well as being somewhat flattened laterally. Along the back there is one dorsal fin which extends the whole length of the animal to its tail. Behind the operculum are small pectoral fins and below these lie a pair of minute spiny pelvic fins. Butterfish are usually a mustardy-brown colour with a row of about eleven dark spots, each with a pale outer ring, running along the length of the body beneath the dorsal fin. Butterfish are common intertidal animals found on the shore in all but the bitterest winter weather, when they tend to migrate for a short while to deeper, more equable water. They are usually found on the rocky shore and in pools but seem to prefer to remain under rocks or seaweed when the tide is out. As their name suggests they are slippery fish. They do have scales but they are tiny and deeply embedded, so their slimy skin and wriggly habit makes them very difficult to pick up.

Facing page: Butterfish are sad-looking creatures often found in the intertidal zone.

BUTTERFISH

A broken shell provides a useful refuge for this butterfish.

Despite their dislike of the cold, butterfish start to breed along the coastline during the winter, usually some time between November and March. They become sexually mature in their second year, when they are over 10cm long. The eggs are laid in a clutch under stones, in a crevice or in an empty bivalve shell. The parents take it in turns to guard the eggs, a task they spend so much time over that for a while they stop feeding. For extra protection it is not uncommon for one of the fish, usually the female, actually to coil its long body around the eggs so that it hides them from predators. After about four weeks the small, 1cm long, larval fish emerge and enter the plankton. Here they remain until about May, when at around 3cm they start to migrate inshore. Butterfish can live for up to five years and have a varied diet of small crustaceans and worms, as well as many other invertebrates.

BUTTERFISH	*Pholis gunnellus*
Size:	Up to 25cm
Colour:	Mottled brown with dark spots along back
Habitat:	Rocky, sandy, muddy seabed; mid-shore to 50m

Blennies are a common rockpool inhabitant.

SHANNY	*Lipophrys pholis*
	(Blennius pholis)
Size:	Up to 15cm
Colour:	Olive green/brown, breeding
	males black
Habitat:	Rocky shore; low shore and
	shallow sea

The alert-looking shanny (sometimes known as the blenny as it is the most common type of fish in the blenny group) is one of Britain's most familiar rocky shore fish. Like many other fish found on or near the shore, all blennies are a shade of yellow or olive green/brown, with varying degrees of dappling or stripes. This coloration camouflages the fish well against their natural background of rock and weed. Shannies' smooth, scaleless skin enables them to slither and squeeze through small cracks and crevices. Along the back is a single dorsal fin which is straight along the whole length. This is in contrast to gobies which have both scales and two dorsal fins.

All blennies, but in particular the shanny, have adapted well to shallow seas, rockpools and life on the shore. Young shannies live high on the shore, moving into deeper water as they grow. All undertake a seaward migration late in the year to escape the bitter cold of the terrestial winter. During the rest of the year they remain in the moist environment under weeds or beneath stones where they can happily cope with being left exposed until the tide returns. They can also withstand wide fluctuations in salinity.

At night it is not unusual to see them climb to the edge of a pool and put their heads out of the water to gulp air. They do this because although seaweeds release oxygen during the day, they use it up at night and deplete the levels to such an extent that any animals in the pool are at risk of suffocation. Many shanny have a favourite pool or crevice to which they return, often for weeks at a time, after feeding expeditions.

SHANNY

Blennies scavenging at high tide.

Shannies swim in upward jerking darts, followed by slow gliding descents back towards the seabed. However, if they are stranded by the tide they can use the strong pectoral fins (which lie behind the pelvic fins) to 'walk' to either the sea or a dark, damp patch of seaweed where they hide till the tide comes back in. Shannies have large eyes which are put to good use in the hunt for food. They are omnivorous in habit and the teeth in both the upper and the lower jaw are small, sharp and closely packed. These strong teeth enable them to consume a diverse range of foods. Young shannies tend to feed exclusively on the feeding appendages of barnacles, while mature fish are able to grind whole barnacles, in addition to other crustaceans and molluscs, to extract the meat.

As with most British fish, breeding takes place from spring to early summer. In the shanny this readiness to mate is accompanied by a colour change in the males, which become dark grey or black with startling white lips. This acts as a signal to the females, telling

The receding tide may leave blennies stranded but they happily await the water's return.

them that they are ready and willing to mate. Each male will entice a female to lay her eggs in batches on a rock ledge, in a shell or crevice. She may lay up to eight clutches, each containing up to 1,000 white eggs. Once the eggs have been fertilized and the disc-shaped base has been securely cemented down, the female loses interest and leaves the male in charge. The eggs are guarded throughout the 6–8 weeks it takes them to hatch. During this time the males work hard to chase away any unwelcome intruders and to maintain the required oxygen levels by fanning fresh seawater over them. By summer the planktonic larvae have become quite common and the young start to settle out on the shore when they are about 2cm long.

Shannies have a relatively short life. They normally become sexually mature in their second year and usually only live about five years athough the longest recorded lifespan of a shanny in captivity is fourteen years.

*The tompot blenny
appears to be quite
inquisitive.*

TOMPOT BLENNY	*Parablennius gattorugine*
Size:	Up to 25cm
Colour:	Olive brown, mottled
Habitat:	Rocky shore and kelp bed;
	low shore and shallow seas
	to 35m

With their large eyes and wide 'smiling' mouths tompot blennies look rather like shannies at first glance. In fact, they differ in a number of ways. First, they grow larger and have an unmistakable pair of branched antennae above each eye. They also have seven or more distinct brown bars running round from the dorsal fin to beneath the body and there is a slight indentation halfway down the dorsal fin, the first part of which is lower and has stiffer fin rays than the second. Unlike shannies they are uncommon on the shore and it is quite rare to find one in a rockpool or under seaweed. They prefer deeper areas down to a

TOMPOT BLENNY

Background: Eggs of the tompot blenny.
Inset: Tompots have an unmistakable pair of branched antennae over each eye which, along with their 'smiling' mouths, give them a comical appearance.

depth of 35m and tend to live hidden in small holes and crevices within rocks or wrecks. Tompot blennies are more common to the south and west of the country and are sometimes caught by trawlermen or in crab or lobster pots, which tends to suggest that they are willing to leave the safety of their holes to go in search of food.

Tompot blennies have a broad diet which includes worms, jellyfish, crustaceans, anemones, hydroids and molluscs, as well as fragments of seaweed. They grow quickly and can live for up to nine years. They become sexually mature in their second or third year and their breeding behaviour is similar to that of shannies. Again it is the male that takes the active role in picking a site for the nest and in guarding the clutch of eggs during the four weeks it takes them to mature and hatch.

G obies are small (usually less than 15cm long) and very abundant fish. At times they can be found together in huge numbers at the seashore and in shallow pools. Worldwide there are about 1,500 different species of goby and fifteen are found in northern European waters. Superficially gobies are similar to blennies. Both are bottom-living fish and have tapering bodies and short blunt heads with large fleshy mouths. They also have their large eyes close to the top of the head so that they can see predators and prey approaching. However, close inspection will show a number of characteristic differences. Gobies have two quite distinct dorsal fins and their pelvic fins (like those of clingfish, lumpsuckers and sea snails) have been modified into fan-shaped suckers. Their lateral line is either much reduced or absent and they have conspicuous scales which, in the rock goby, have extended to form a sharp spine at the top of each pectoral fin.

Because gobies look so similar a knowledge of their precise habitat is helpful in distinguishing one species from another. They are usually very particular about their choice of locality. The two-spot goby lives in small shoals among weed, well off the seabed. Sand gobies are abundant in intertidal pools but shun similar locations in estuaries and saltings, where the common goby is found, because it is not so tolerant of fluctuations in salinity. Black gobies are almost entirely restricted to estuaries and sheltered bays, enjoying muddy areas where the eel grass *Zostera* is found. In contrast the rock goby, as its name suggests, is restricted to intertidal and shallow rocky shores.

All gobies are shades of dapple brown, which camouflages them well against the coastal scenery. The common and sand gobies are especially well coloured with a mottled sandy skin and occasional darker markings along their sides, together with fainter brown marks on the fins which help to break up their outline. They are also able to hide themselves beneath the sand or mud by a quick flick of the fins which buries them, leaving only their observant eyes exposed. In this position whole shoals remain motionless so as not to catch the eye of predators like little terns, while they themselves may suddenly dart out to catch their prey.

Living as they do at the edge of the seashore with the continuous motion of the currents, waves and tides, gobies are always in danger of being swept away from the shoal. It is in this situation that their sucker is used. Although the strength with which it attaches is weak compared to that of a lumpsucker or clingfish, it is perfectly able to hold the fish down to a rock or pebble in normal conditions.

Gobies eat a variety of different types of prey. The specific type depends on the environment in which they live. The open-water two-spot goby is known to eat small crustaceans such as copepods and mysids

SAND GOBY	*Pomatoschistus minutus*
Size:	Up to 11cm
Colour:	Dapple sandy brown
Habitat:	Sandy seashore; low shore and shallow sea to 40m

COMMON GOBY	*Pomatoschistus microps*
Size:	Up to 7cm
Colour:	Dapple grey
Habitat:	Sandy and muddy seabed and estuaries; low shore and shallow sea

Facing page: Sand gobies may be mistaken for weever fish, to which they bear a superficial resemblance.

GOBIES

The rock goby is restricted to the intertidal zone and shallow, rocky shores.

ROCK GOBY	Gobius paganellus
Size:	Up to 12cm
Colour:	Dark dapple rusty brown, first dorsal fin has pale yellow (female) or orange (male) band
Habitat:	Rocky seashore; low shore and shallow sea to 10m

TWO-SPOT GOBY	Gobiusculus flavescens
Size:	Up to 6cm
Colour:	Dapple rusty brown, dark spot on first dorsal fin and base of tail
Habitat:	Rocky seashore among weeds; shallow sea to 15m

BLACK GOBY	Gobius niger
Size:	Up to 17cm
Colour:	Dark dapple brown
Habitat:	Sandy and muddy seabed; low shore to 70m

Rock gobies may excavate under rocks to form a refuge.

and the eggs individually attached at the base. Once her work is over, the female leaves and the male is left to attend to the eggs by warding off predators and by aerating and cleaning them throughout the 2–3 weeks they take to hatch. As they mature the dark eye rings of the larvae can clearly be seen through the outer casing. Once hatched the larvae go through a short planktonic phase until, at about 12mm long, they join the main population on the sea floor.

Gobies are short-lived fish. They become sexually mature in their first year and rarely live longer than two years. Only the larger black goby is thought to mature later and perhaps reach an age of four or five years.

as well as larval fish, while the bottom-living gobies take brown shrimps and the larvae of mussels, cockles and flatfish.

All gobies spawn in spring and summer. At this time the breeding males may display a number of secondary sexual characteristics which distinguish them from the females. They usually become more pigmented and brightly coloured and their dorsal fin rays are often raised higher. Both male and female fish may have a number of partners during the season, but first the male must attract the attention of a receptive female. Once he has done this he leads her to a suitable egg-laying site. Usually this is an empty bivalve shell or a clean piece of rock. With the two-spot goby the hollow holdfast of the kelp *Saccorhiza polyschides* is popular and it is not unknown for empty worm casts to be chosen. Once she is happy with the site the female will then lay her clutch of eggs. Rock gobies lay sheets of eggs while in the other species clutches can be vase- or pear-shaped

A close-up view of the rock goby's head.

GREY GURNARD	*Eutrigla gurnardus*
Size:	Up to 45cm
Colour:	Variable; dapple grey/brown, cream below
Habitat:	Sandy, muddy seabeds; shallow sea to 200m

RED GURNARD	*Aspitrigla cuculus*
Size:	Up to 30cm
Colour:	Rosy orange/red, pale below
Habitat:	Sandy, gravel and muddy seabeds; shallow seas from 20m to 250m

TUB GURNARD	*Trigla lucerna*
Size:	Up to 60cm
Colour:	Dull red with brilliant blue edge to pectoral fins, pale beneath
Habitat:	Sandy, gravel and muddy seabeds; shallow seas to 200m, most common at 50–150m

Gurnards are distinctive fish with large heads and tapering bodies. There are seven species found in British waters, three of which, the red, grey and tub gurnards, are very common.

Their skin is very tough and the head is covered in an armour of strong, dermal, bony plates. In addition, there are sharp spines of varying length on the gill coverings. Gurnards are brightly coloured fish on their upper surfaces and paler below. Tub gurnards in particular are remarkable in that although the body is a dull browny red colour, their red pectoral fins have an iridescent peacock blue fringe.

An unusual feature of gurnards is that the pectoral fin has been modified in such a way that the first three fin rays have separated out into finger-like projections. Using these projections, which are covered in sense cells, the fish are able to explore the seabed for food. They also use them to walk delicately over the sea floor and to raise themselves up to get a better view of the area.

Gurnards live together in small, loosely knit shoals over soft substrates such as sand, mud or gravel. They are common both inshore and in deeper water.

Grey gurnards occur around the whole of Britain's coastline, while red and tub gurnards are more usually found to the south and west of the country. They are rather noisy creatures, producing loud grunting sounds. They make these grunts by the contraction of muscles acting on the swim bladder and causing it to vibrate. These noises are thought to keep the shoal together as they swim over or explore the seabed.

Feeding both by day, when their eyesight is important, and by night, when the fin rays detect the food, gurnards mainly eat bottom-living animals. All of them show a preference for crabs, shrimps, prawns and squat lobsters. Grey and tub gurnards are also known to eat fish, including sand eels, flatfish, gobies and sprats, and tub gurnards will sometimes also take worms, fish eggs, small cockles, scallops and other bivalve molluscs.

Gurnards can live for 6–8 years but they are only three or four years old when they become sexually mature. Each year they move inshore to spawn. Spawning takes place in early summer from April to August for the grey gurnard and from May to June for the tub gurnard. All gurnards lay small eggs (about 1.5mm in

diameter) which float freely in the sea. After a short period of 8–10 days they hatch and the young larvae spend the next 6–10 weeks feeding off zooplankton in coastal waters or even, as in the case of the tub gurnard, in shallow bays and estuaries. By this time they have grown to about 3cm and they migrate down to the seabed where, like their parents, they group together into small shoals.

Facing page: Gurnards 'walk' across the seabed by means of their modified pectoral fins.

This page: Grey gurnards are known to eat fish as well as a variety of crustaceans.

GURNARD

These fish get their name from the long spines which project out from in front of the operculum. Each spine is longer than the diameter of the eye. Even though they are commonly called sea scorpions, they have no venom and there is no danger at all in gently handling them. Sea scorpions are usually found in a similar habitat to common blennies. They live in shallow seas and among the rocks and brown seaweeds of the intertidal zone, where their dappled greeny brown skin means that they are well camouflaged. They have large eyes and mouths and are visual predators, hunting during the day for small fish and invertebrate animals.

When they are about two years old and over 10cm long sea scorpions become sexually active. Breeding generally takes place between January and May when groups of orange-coloured eggs are laid under rocks or seaweed or in crevices on the shore. After six or seven weeks, the eggs hatch into planktonic larvae, and it is not until they have grown to just over 1cm that they become bottom-living.

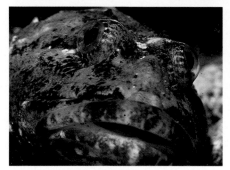

LONG-SPINED SEA SCORPION

This page and facing page: The well-camouflaged sea scorpion is actually harmless in spite of its name.

LONG-SPINED SEA SCORPION	
Taurulus bubalis (Cottus bubalis)	
Size:	Up to 18cm
Colour:	Olive brown/green
Habitat:	Rocky shores; mid-shore to 30m

LUMPSUCKER	*Cyclopterus lumpus*
Size:	Females up to 60cm, males to 50cm
Colour:	Grey/blue, male red/purple when breeding, young olive/green
Habitat:	Rocky seabed; very low shore to 300m

Above and right: The lumpsucker is another distinctively shaped fish. The cream-coloured sucker beneath the head is used very effectively to cling to rocks.

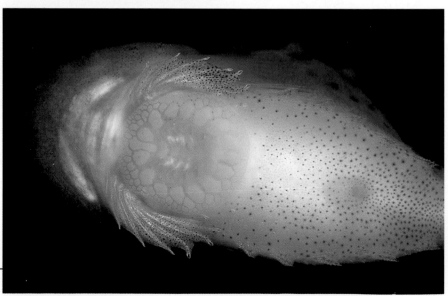

There is only one species of lumpsucker found around Britain and it has a quite distinctive shape, with a thick body, scaleless skin and rounded head. The fish are usually a dull slate grey/blue in colour which matches their rocky surroundings. Only when the males are in breeding condition do they take on a purple/rusty hue. Beneath the head is the pale cream-coloured sucker from which the fish gets its name, and which is used for clinging to rocks. This structure has developed from the fusion of the pelvic fins and is very effective. Only occasionally, in extremely stormy weather, will the fish be wrenched from the rocks by the ferocity of the waves and thrown on to the beach to die of exposure.

Although scaleless, the skin of the lumpsucker is protected by a covering of bony denticles and seven rows of sharply pointed tubercles, three along each side and one down the back. The young fish have two dorsal fins, but as they grow the first, fleshier one disappears. Females tend to have a larger dorsal 'hump' but smaller pectoral fins than the males. Both sexes suffer badly from fish lice. The small parasitic crustaceans embed their heads into the skin and suck away at the flesh, rarely killing their host but no doubt causing considerable irritation.

Lumpsuckers, which have no swim bladder, are mainly bottom-living fish found among firm rocky substrates. The young may be quite agile and can swim rapidly in short bursts, but the adults are slow-moving and stop for regular rests, using their suckers for attachment. Their diet appears to be quite varied, and is known to include small crustaceans, soft-bodied molluscs, worms, comb-jellies and small fish. Lumpsuckers are caught in the commercial catch and although they are not commonly eaten in Britain they are well liked in northern Europe, where they are sold either fresh, smoked or salted. Even more popular are the large eggs (roe). These are treated with salt and at times dyed black before they are sold as 'lumpfish caviar' – a reasonable alternative to real sturgeon caviar.

In spring lumpsuckers move from deeper water towards the shore. The females find a suitable rocky ledge upon which they will lay up to a quarter of a million grey to red eggs, after which they soon migrate offshore again. It is the male that remains behind to guard the eggs attentively, never leaving them throughout the 6–7 weeks it takes for them to mature to an opaque green colour, at which time they are ready to hatch. He does not feed at all during this period, but spends his time clamped on a rock close to the egg mass and fans the eggs or blows water over them to keep them aerated and free of silt. He will also chase away any predators or even physically remove threatening creatures like crabs or starfish.

At times, in the far north of Scotland, the eggs are actually laid intertidally. It is in places like this, during low-tide, that they are at the greatest risk from predators. The male remains close, attached to a nearby rock, but he is helpless to do anything against terrestial predators such as rats and birds. Not only will these predators take the eggs but he is himself very vunerable to attack. By the time the eggs hatch and the larvae disperse, the male, if he has survived, is usually exhausted.

Lumpsuckers usually spend their first summer among the kelp forests where they can change their colour to match their background. The suction disc is already formed when they hatch and so they are able to cling to the undersides of weeds or rocks, wrap their tails around their heads and 'play possum', looking just like blobs of jelly. They are quite slow-growing fish, reaching 5cm after the first year, sexual maturity by the third and full size at about five years of age. They do not appear to have many predators, but they are occasionally eaten by dogfish, anglerfish and seals.

LUMPSUCKER

The black sea bream is delicious to eat and is much sought after by anglers.

Black sea bream, sometimes known as old wife, are deep-bodied, silver-scaled fish which are popular with anglers. Although delicious to eat, they are rarely found in the fishmongers because they live close to the shore near or among rocky faces, and as a result they are hardly ever caught by professional trawlermen. They are omnivourous, and in addition to eating small fish and crustaceans, they also nibble at the algae and small animals that encrust the coastal rocks.

Sea bream are most often found in the English Channel and along western coasts. They seem to come from Spain and the south-west of France to feed and breed. They spawn, in April and May, in an area of the Channel between Portland Bill and Brighton and have a complex courtship. In spring the males lose their overall grey metallic sheen and develop six or seven vertical black and silver bands. Each fish spends a considerable amount of time forming a nest by creating an indentation in the gravel bed, which

he then fiercely guards. When a female starts to take an interest in a male, he leads her to his nest and entices her to lay her eggs, which are white and about 1mm in diameter. Once the eggs are glued firmly to the gravel base of the nest he fertilizes them and guards them until they hatch. At hatching time certain areas around the Isle of Wight are often thick with tiny bream. From here many continue to move north around the west coast of Britain, occasionaly as far as southern Scotland.

BLACK SEA BREAM

BLACK SEA BREAM *Spondyliosoma cantharus*	
Size:	Up to 50cm
Colour:	Silvery grey, at times with dark vertical bands
Habitat:	Open sea; coastal waters around the southern half of Britain

The camouflage of the anglerfish is superb, allowing it to bait and catch its prey.

Anglerfish are extraordinary to look at, with many unusual features. The head is huge and has a wide, gaping mouth with four rows of sharp, inward-pointing teeth. The lower jaw is fringed with a series of fleshy barbels which may be up to 5cm long and look very like bits of straggly seaweed. On top of the head there are three dorsal spines, the first of which has a well-developed fleshy tip made up of a little tuft of filaments and it is this which is used to bait prey. All this, combined with the fact that the scaleless skin is loose and limp, gives the animal a somewhat grotesque look.

In fact it is superbly adapted to life as a hunter on the sea floor. Growing large – up to 2m within a few years – it is a voracious predator and has a fascinating technique for capturing its food. It is by nature a slow swimmer, so it cannot chase its prey. Instead it excavates a small hollow in the sand or mud of the seabed and the debris from this excavation clouds the water, obscuring all but the lure on the first dorsal spine. Other fish will notice this fleshy flap as it twitches about. As a fish moves in to investigate the possibility that it may have found some food, the anglerfish will wait until it is close enough and then skilfully flick the lure down, out of reach, while at the same time catching the prey with one snap of its jaws. They are known to catch and eat a wide variety of fish in this way, including rays, sand eels, flatfish, members of the cod family and at times conger eels. They also take bottom-living invertebrates such as lobsters and crabs. In addition, although they usually live in fairly deep water, they sometimes come close enough inshore to catch water birds, including duck and coot, and on at least one occasion, an anglerfish

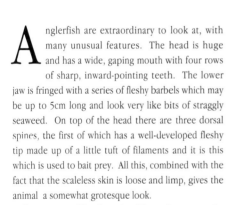

ANGLERFISH	*Lophius piscatorius*
Size:	Up to 2.0m
Colour:	Dark olive brown above, white below
Habitat:	Sandy and muddy seabed; shallow sea to 550m, to 1,800m when spawning

ANGLERFISH

has been found dead having choked on a seagull! When spawning activity starts in spring, anglerfish migrate to deeper water (as deep as 1,800m) to the west of Britain. They become sexually mature at about 75cm and will spawn any time between March and July. They shed their eggs, not individually, but in large rafts of violet-coloured, gelatinous material. Each ribbon-like structure may be as much as 9m long and 1m wide and will hold up to a million eggs in a single, or occasionally a double, row. Drifting freely in the sea, this mass of eggs is soon broken up by the water's movement, allowing the small straw-coloured eggs to be distributed over a wide area.

When the young first hatch they are only about 4.5mm long and look somewhat different from their parents. They are well rounded and have large wing-like fins together with dorsal spines that are drawn out into long delicate projections. These features help them to remain bouyant and drift freely in the sea. The young planktonic anglerfish grow quickly. They have the same large appetite as their parents and eat huge numbers of small crustaceans and arrow worms. As they grow their body gradually takes on the adult form and by the time they are 6–8cm long they migrate down to areas of shallow seabed and take on the adult lifestyle.

This voracious predator sometimes comes close enough to the shore to catch water birds.

Above: Turbot boast superb patterning.

Right: It is clear that turbot are at home on sand.

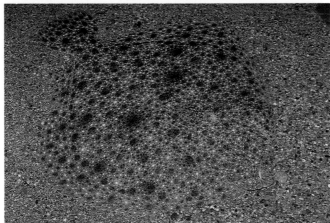

TURBOT	*Scophthalmus maximus*
Size:	Up to 100cm, no scales on eye side
Colour:	Variable, sandy brown
Habitat:	Sandy, muddy seabed; shallow sea to 80m

F latfish is a general term given to those familiar fish such as plaice in which the body is flat, asymmetrical, thin and leaf-like. There are 24 different species of flatfish in the seas around Europe, many of which are caught in trawls or seine nets, or by hook and line, for their delicious, highly prized flesh.

While flatfish can be divided into three seperate groups – the turbot family, Bothidae, the plaice family, Pleuronectidae, and the sole family, Soleidae – there are many characteristics which they all share. The most obvious is that their eyes are both on the upper side of the head. Fish whose left side is uppermost and whose right eye has migrated so that both eyes are to the right of the mouth (turbot, brill and topknot) are called left-eyed flatfish. Those whose

FLATFISH

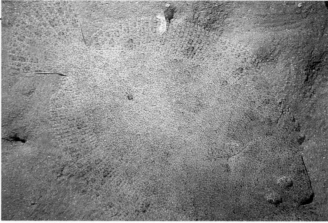

right side is uppermost, so that the eyes lie to the left of the mouth (plaice, flounder, sole and dab), are called right-eyed flatfish.

In fact, it is not uncommon for a fish of a given species to be found with the eyes on the 'wrong' side of the mouth. This occurs during the early development of the young fish which, when they hatch out, look just like any other larval fish. They are cod-like in shape and have one eye on either side of the head. The larvae grow quite normally for a number of days until they suddenly keel over. In the Dover sole for example, this happens some time between the eleventh and thirteenth days and they keel over to the left. As a result the right-hand side of the body becomes the upper surface and the left eye lies below. It is at this time that the young fish takes up residence on the seabed and so all the lower eye would be able to see would be the soft substrate of the sea floor. To overcome this, part of the larval metamorphosis includes the migration of the eye from the lower to the upper side of the body. This means that for the rest of its life the fish will lie on its blind side, while the eyes on the upper surface can be used to their full advantage.

This flattening out on to one side of the body is radical and leads not only to the displacement of the eyes but also to most of the rest of the body's organs and appendages. The fish are totally asymmetrical. The brain and the bones surrounding it are offset. The gills and their covering, the operculum, lie beyond the mouth away from the eyes. They have a special adaptation which enables water taken in by

Above left: Only the eye gives the face of this brill away.

Above right: The brill is well camouflaged in its sandy habitat.

BRILL	*Scophthalmus rhombus*
Size:	Up to 60cm, with small, smooth scales
Colour:	Variable; sandy grey/brown
Habitat:	Sandy, muddy seabed; shallow sea to 75m

TOPKNOT	*Zeugopterus punctatus*
Size:	Up to 25cm
Colour:	Variable; dapple rusty brown
Habitat:	Rocky seabed; low shore and shallow sea to 40m

Flounder.

The flounder makes use of tidal currents in its long seasonal migrations.

the mouth to pass over both gills before being diverted out, by a special channel, on to the lower side. The guts lie on the same side as the mouth. Finally, the jaw on the side away from the eyes is far more robust, has stronger teeth and is better developed than that on the upper side.

Another remarkable change that takes place during metamorphosis is that pigment cells only function on the upper surface of the flattened fish. As a result flatfish are normally white on their underside and the zig-zag shaped blocks of muscle and blood vessels can be seen quite clearly. The upper, pigmented, surface is completely different. It is variable in colour, both between different species of fish, depending on whether they live on sand, rock, etc., and also between fish of the same species, depending on where they are. Not only can they bury themselves by flicking sand over their surface but they can also change colour to suit their background. It is a remarkable camouflage and enables them to avoid predators and also to remain unobserved by their prey. The colour changes that take place are controlled by nerves and hormones which contract or relax the pigment cells. Each pigment cell (chromatophore) is like a bag of coloured granules which when contracted is just a small dot of brown (making the fish look pale) but when relaxed is a large blotch (making it look dark). In the plaice this mechanism is so good that when the fish is put on a chess board it will actually display dark and light squares over its body! Not suprisingly for such a complex system, mistakes do happen and at times albino fish are found, as are fish with patches of colour on their undersides.

Most flatfish live on soft substrates; only the topknots have adapted to a life among rocks and kelp, to which they can cling tightly. They are bottom-living and do not have a swim bladder. Their range extends from the edge of the continental shelf

DAB	*Limanda limanda*
Size:	Up to 40cm
Colour:	Variable; sandy brown with dark spots
Habitat:	Sandy, muddy seabed; shallow sea to 150m

PLAICE	*Pleuronectes platessa*
Size:	Up to 90cm
Colour:	Variable; sandy brown with bright orange spots
Habitat:	Sandy, muddy and gravel sea-bed; shallow sea to 300m

to shallow coastal waters. Young flatfish can be found in huge numbers during the summer in the warm water at the edge of sandy seashores. Here they are often caught in shrimp and sand-eel nets and may be tiny – no bigger than a 10p piece. Many flatfish are able to live at the mouths of estuaries where there is a reduction in the salinity of the seawater. Flounders are the most adaptable and can actually move up the estuary and live in freshwater for short periods.

Many flatfish undertake long seasonal or breeding migrations, usually at night. During the autumn they may move away from cold inshore areas to deeper water where they will remain, feeding little if at all. In spring they will move back inshore again to feed and spawn.

The plaice population is divided into a number of local races, each differing in the number of fin rays or vertebrae they have. Each also has quite specific spawning grounds to which they return every year, the principal British grounds lying to the east of the country on the Great Fisher Bank. Flounders also undertake long migrations and it is known that they can continue to move in a given direction for hours at a time. To make their journey a little easier they make use of tidal currents, moving up towards the surface to swim along with the inflooding tide and going back down to the seabed where it is easier to continue against the ebb tide.

As has been said, many of the flatfish migrate to specific locations when they are about to spawn. The mature fish are known to congregate together in large shoals in spring or early summer and this helps to ensure that as many of the eggs as possible are successfully fertilized. Because the vast majority of eggs and young larvae die or are eaten during the first days of life, huge numbers of eggs are released. A female plaice can lay up to half a million, while each female turbot may release between ten and fifteen million. The eggs have a small oil droplet that ensures that they will float freely just below the sea surface. The currents then carry them away from the breeding grounds and disperse them in areas where there will be less competition with the adults. The eggs hatch after only a few days and the larvae grow rapidly, eating their fill to give them the strength they need to undertake the amazing metamorphosis that prepares them for life on the seabed.

In common with other flatfish, the sole has both eyes on the same side of its head.

FLOUNDER	*Platichthys flesus*
Size:	Up to 20cm
Colour:	Variable; greeny/brown with brown or dull orange spots
Habitat:	Sandy, muddy seabed and estuary; shallow sea to 50m

DOVER SOLE	*Solea solea*
Size:	Up to 50cm
Colour:	Variable; dapple grey brown
Habitat:	Sandy, muddy seabed and estuary mouth; shallow sea to 200m

Antenna — A sense organ – usually a long feeler on an animal's head.

Benthic — Living in or on the seabed.

Brackish water — Water which is less salty than normal seawater.

Byssus — Small fibres made by the feet of bivalve molluscs. They are used for attachment.

Crustacean — Class of arthropods with two pairs of antennae, a hard outer skeleton and jointed limbs. Over 1,850 species are known in Britain, including crabs and lobsters.

Denticle — Small, closely packed, tooth-like structures which protect the skin of sharks and rays.

Detritus — Decomposing plant and animal material.

Dorsal — The upper part or back of an animal.

Ecology — The inter-relationship between an organism or group of organisms and their environment.

Environment — The surroundings in which an animal or plant lives.

Fin — A group of spines linked by a membrane which projects out from the body of a fish. It helps to propel and stabilize the fish.

Filter feeder — An animal that obtains its food by sieving it from the surrounding water.

Fusion — Joining together.

Gills — The breathing apparatus of fish.

Gill slits — The gaps between the gills through which water flows.

Gonad — Reproductive organ of an animal.

Habitat — An area with certain physical characteristics which supports a particular community of plants and animals.

Herbivore — An animal which eats plant material.

Hermaphrodite — An animal that functions as both male and female.

Holdfast — The attachment device of seaweeds.

Intertidal — The area of the shore between high- and low-water spring tides.

Invertebrate — Animal without a backbone.

Larva — The immature, pre-adult or juvenile stage of an animal. Often active or free-living and therefore an important dispersal phase.

Lateral — To the side.

Lateral line — A line of pressure-sensitive cells lying in a perforated canal along the side of a fish. The line may look either dark or light.

Metamorphosis — The dramatic change in form and structure which occurs when a larva changes into an adult.

Migration — Regular movements of animals between feeding, breeding or wintering grounds.

Milt — The spawn of male fish, the soft roe.

Mollusc — Phylum of invertebrate animals with soft bodies and often a chalky protective shell; they include mussels, whelks and octopus.

Moult — The shedding of the outer skeleton associated with growth in crustaceans.

Neap tides — The tides with the smallest difference in height between high and low water.

GLOSSARY

Omnivore   An animal that eats both plants and animals.

Operculum   The gill cover of a fish.

Parasite   An organism that lives on and gets its food from another organism (the host).

Pheromone   A chemical released externally by an animal which affects the behaviour of other members of the same species.

Photosynthesis   The process by which green plants make organic compounds from water and carbon dioxide using energy from the sun.

Phylum   A major division of the animal kingdom which includes animals thought to have a common evolutionary origin.

Plankton   Small plants and animals that live, drifting passively, in the surface layers of the sea.

Polyp   A sedentary animal with a sac-like body with one opening, the mouth, surrounded by tentacles.

Predator   An animal that forages for live food.

Prey   An animal which is hunted for and eaten by a predator.

Protrusible   An organ that can extend out from the body of an animal.

Roe   Term used to describe the mass of eggs in a female fish.

Salinity   The salt content of water.

Scales   Flat, rigid or flexible plates that form a protective covering over a fish.

Sedentary   Bottom-living, sitting on the seabed.

Spawning   Laying and fertilizing eggs.

Species   A biological group, or population, which can interbreed freely together but not with others.

Spicule   Tiny piece or fragment of skeletal material.

Spiracle   Small opening in the head or neck of primitive fish.

Spring tides   The tides with the largest difference in height between high and low water.

Stipe   The stalk-like part of a seaweed.

Swim bladder   The gas or air-filled sac which lies between the gut and backbone in teleost fish.

Teleost   Bony fish.

Terrestial   Land-based.

Territorial behaviour   The behaviour of an animal that sets up, maintains and defends a breeding site, home range or feeding area.

Tide   The periodic rise and fall of sea level.

Tubercle   Small benign rounded swelling.

Ventral   The underside of a plant or animal.

Vertebrate   An animal with a backbone.

Zooplankton   Animals found in the plankton.

Zonation   The division of the shoreline into horizontal bands with a characteristic fauna and flora.

Dicentrarchus labrax, 146
Dog whelk, 14
Dogfish, 122-125, 132, 173
Dover sole, 178, 181
Dragonet, 123, 155

Echinocardium cordatum, 22, 121
Echinodermata, 118
Echinus esculentus, 118
Edible crab, 93-94, 98
Edible sea urchin, 118-119
Eel, 25, 124, 131-132, 141, 144, 157, 164, 169, 175
Elephant's ear sponge, 43
Entelurus aequoreus, 142
Estuaries, 23-25, 27, 29-30, 32, 35, 47, 95, 141-142, 146, 164, 169, 181
Eutrigula gurnardus, 168

Feather star, 15, 103
Flatfish, 25, 29, 123, 129-130, 132, 146, 154, 167, 169, 175, 177-178, 180-181
Flounder, 178-181
Fucoid weed, 13

Gadoid, 130, 144
Gadus morhua, 136
Galathea squamifera, 80-81

Galathea strigosa, 80
Gobius niger, 166
Gobius paganellus, 166
Gobiusculus flavescens, 166
Goby, 17, 136, 154, 159, 164-167, 169
Goldsinny wrasse, 153
Grantia compressa, 43
Greater pipefish, 141-142
Green sea urchin, 117
Grey gurnard, 168-169
Gunnel, 157
Gurnard, 123, 168-169

Halichondria bowerbanki, 43
Halichondria panicea, 43
Heart urchin, 121
Helcion pellucidum, 20
Henricia oculata, 108
Hermit crab, 25, 42-43, 56, 81-84, 130
Herring, 123, 132, 135-136, 146
Holdfast, 19-20, 167
Holothuria forskali, 27
Homarus gammarus, 76
Hyas araneus, 88
Hydrobia ulvae, 27
Hydroid, 83, 85, 108, 163

Inachus dorsettensis, 89

Jellyfish, 5, 44-49, 51, 55-59, 61, 63, 65, 138, 163
Jewel anemone, 6, 61-62
John dory, 144

Kelp, 15-16, 18-20, 50-51, 73, 81, 94, 101, 103, 113, 118, 141, 144, 162, 167, 173, 180

Labrus bergylta, 150
Labrus mixtus, 151
Laminaria digitata, 20
Laminaria saccharina, 19
Lanice conchilega, 25
Leptocephalus, 132
Lice, 173
Limanda limanda, 180
Limpet, 17, 20, 67
Liocarcinus puber, 100
Lipophrys pholis, 159
Lobster, 5, 73, 75-82, 88-89, 91, 93-96, 99-101, 123, 132, 149, 163, 169, 175
Long shore drift, 21
Long-legged spider crab, 85
Long-spined sea scorpion, 170-171
Lophius piscatorius, 175
Luidia ciliaris, 104-105

Lumpsucker, 164, 172-173

Macropipus depurator, 98
Macropodia rostrata, 85
Maja squinado, 86
Marthasterias glacialis, 112
Masked crab, 91
Membranipora membranacea, 20
Merlangius merlangus, 139
Mermaid's purse, 124-125, 130
Metridium senile, 55
Mollusc, 5, 13, 23, 35, 67-69, 71-72, 75, 97, 112, 117, 123,
127, 151, 155, 160, 163, 169, 173
Muddy shores, 27
Mullet, 27, 29
Mussel, 14, 32, 55, 66-67, 69, 93, 97, 109, 111, 167
Mustelus asterias, 126
Mustelus mustelus, 126
Mysid, 29, 164
Mytilus edulis, 66

Neap tide, 27, 29-30
Neomysis integer, 29
Nereis fucata, 83
Nerophis lumbriciformis, 143
Nurse hound, 123

Octopus, 39, 44-45, 67, 70-72, 132
Old wife, 174
Ophiothrix fragilis, 113

Opossum shrimp, 29
Oyster, 93, 109, 111

Pachymatisma johnstonia, 43
Pagurus bernhardus, 56, 82
Palaemon serratus, 74
Palinurus vulgaris, 75
Parablennius gattorugine, 162
Paracentrotus lividus, 116
Parasitic anemone, 56
Peacock worm, 26
Pedicellariae, 112, 118
Peltogaster paguri, 83
Periwinkle, 14
Phallusia mammillata, 59
Pholis gunnellus, 158
Phragmites, 29
Piddock, 13
Pilchard, 123, 144
Pipefish, 13, 141-143
Plaice, 177-178, 180-181
Platichthys flesus, 181
Pleuronectes platessa, 180
Plumose anemones, 54-55
Pollachius pollachius, 136
Pollachius virens, 139
Pollack, 132, 136, 138
Pomatoschistus minutus, 164
Pouting, 123, 135
Prawn, 7, 17, 49, 53, 73-74, 81, 123, 169

Psammechinus miliaris, 117
Purse sponge, 43

Queen scallop, 68-69
Queenies, 69

Ragworm, 83, 130
Raja clavata, 128
Ray, 25, 109, 118, 128-130, 135, 162, 167, 169, 175, 181
Razor shell, 23, 35
Red gurnard, 168
Rhizostoma octopus, 44-45
Rock goby, 164, 166-167
Rock salmon, 124
Rockpool, 7, 9, 16-17, 49, 53, 74, 80, 95-96, 101, 107, 111,
124, 131, 159, 162
Rocky shore, 12-15, 30, 32, 49, 53, 66-67, 73-74, 80, 82,
86, 94, 96, 103, 109, 117-118, 122, 124, 143, 157, 159, 162,
164, 166, 171
Roker, 129

Sabella pavonina, 25
Saccorhiza polyschides, 167
Sacculina, 97
Sagartia elegans, 57
Saithe, 135, 139
Salmon, 27, 29, 49, 124
Salt marsh, 29
Sand eel, 25, 144, 169, 175
Sand goby, 164-165

Sand mason, 25

Sand, 21-23, 25-26, 32, 48, 85, 91, 93, 98-99, 105-106, 109, 113, 121, 141-142, 144, 154-155, 164-165, 169, 175, 177, 180

Sand smelt, 144

Scallop, 68-69, 169

Scophthalmus maximus, 177

Scophthalmus rhombus, 178

Scorpion spider crab, 89-90

Scyliorhinus caniculus, 122

Scyliorhinus stellaris, 123

Scyphistoma, 46

Sea cucumber, 27

Sea hare, 17

Sea lemon, 17

Sea mat, 20

Sea mouse, 27

Sea orange, 42-43

Sea potato, 22-23, 121

Sea snail, 164

Sea squirt, 41, 59, 117

Sea urchin, 42, 116-119

Shanny, 159-163

Shelduck, 27

Shingle, 21

Shore crab, 95-97

Shrimp, 25, 29, 73, 123, 130, 136, 154, 167, 169, 181

Smooth hound, 126-127

Snake pipefish, 142-143

Snakelocks anemone, 50-51

Solea solea, 181

Spartina anglica, 29

Spider crab, 85-86, 88-90, 99

Spiny spider crab, 85-86, 89

Spiny starfish, 108, 112

Spiracle, 129

Spondyliosoma cantharus, 174

Sponge, 5-6, 15, 17, 41-43, 63, 83, 85, 89-90, 108, 115

Sprat, 130, 144, 146, 169

Spring tide, 15, 20, 25, 27, 30, 113

Squat lobster, 80-81, 169

Squid, 39, 71-72

Starfish, 5, 15, 55, 69, 103, 105-118, 121, 173

Stellate smooth hound, 126-127

Stickleback, 29

Suberites domuncula, 43

Sun star, 109

Swimming crab, 95, 98-101

Syngnathus acus, 141

Taurulus bubalis, 171

Tellin, 23

Temperature extremes, 34-35

Thornback ray, 128-130

Tides, 15, 20, 23, 25, 27, 29-30, 32, 51, 53, 113, 131, 164

Tompot blenny, 162-163

Topknot, 177-178

Trachinus vipera, 154

Trigger fish, 75, 148-149

Trigla lucerna, 168

Trisopterus luscus, 135

Tsunami, 32

Tub gurnard, 168-169

Turbot, 177, 181

Two-spot goby, 164, 166-167

Urchin, 5, 42, 103, 105-109, 111-119, 121

Urticina eques, 53

Urticina felina, 53

Velvet swimming crab, 100-101

Viviparous blenny, 35

Waves, 7, 13-14, 21, 25, 31-32, 67, 116, 146, 164, 173

Weever fish, 25, 154, 164

Whiting, 123, 135-136, 138-139

Worm pipefish, 13, 141, 143

Worms, 25-27, 42, 49, 77, 97, 106, 136, 146, 155, 158, 163, 169, 173, 176

Wrasse, 20, 150-153

Zeugopterus punctatus, 178

Zeus faber, 144

Zooxanthellae, 51

Zostera marina, 25